What is office automation?

How do you plan for and implement successful office automation?

What tools and approaches should b used?

What are the opportunities and risks how is office automation cost justifie

D1622975

Office Automation: A Manager's Guide for Improved Productivity answers these questions by introducing readers to a total systems concept of what office automation means from a practical standpoint. This "how-to" book is written for both non-technical and technical managers, including those with no prior experience in office automation. It provides a framework that helps managers develop methodologies and appropriate plans of action, detailing all the steps involved in making the transition to the office of the future.

The authors familiarize readers with the techniques, tactics, relevant technologies, and strategies necessary to conceptualize and implement successful office systems. Illustrating examples based on their cumulative experiences, they demonstrate how to quantify and qualify the subject...analyze user requirements...and organize and sell an office automation program to senior management. Those issues of office automation that readers must be concerned with in the future are also treated.

Written in an easy-to-follow, concise language, this book features simple methodologies and structural guidelines geared to the macro or micro administrative needs of the user's community. Each chapter of the book maps out the specifics for each of the major areas of office automation. This format allows the novice to learn the entire process, while allowing the expert to concentrate on a particular phase or topic.

Original questionnaires, logs, checklists and other measurement devices included in this hands-on guide will aid all managers in both long- and short-term planning. The authors' more than 40 years of combined experience as managers and professionals in such areas as management information systems, corporate planning, telecommunications, data processing as well as office automation makes this an invaluable working reference.

Office Automation

Office Automation

A MANAGER'S GUIDE
FOR IMPROVED PRODUCTIVITY

Mark A. Lieberman
Gad J. Selig
John J. Walsh

1807 1982

A Wiley-Interscience Publication
JOHN WILEY & SONS
New York • Chichester • Brisbane • Toronto • Singapore

Library of Congress Cataloging in Publication Data:
Lieberman, Mark A., 1944–
  Office automation.

  "A Wiley-Interscience publication".
  Bibliography: p.
  Includes index.
  1. Office practice—Automation. I. Selig, Gad J.,
1939–   . II. Walsh, John J. (John Joseph),
1939–   . III. Title.
HF5548.2.L527   651.8′4   81–23114
ISBN 0–471–07983–9   AACR2

Printed in the United States of America

10 9 8 7 6 5 4 3 2 1

There is nothing more difficult to take in hand, more perilous to conduct, or more uncertain in its success, than to take the lead in the introduction of a new order of things, because the innovator has for enemies all those who have done well under the old conditions, and lukewarm defenders in those who may do well under the new.

Nicolo Machiavelli
The Prince and Discourses

Foreword

When Mark Lieberman, Gad Selig, and Jack Walsh first began working in the office automation field, it did not even have a name. The efforts of these three in defining, analyzing, and building systems their organizations are now using have helped their companies to be on the leading edge in planning for and reaping the benefits of office automation. They have been influential in the explosive growth of the field, and we have shared much information and knowledge with them.

The increasing reliance on technologies in modern business organizations has made it ever more important to be able to plan for and manage technological change. Information processing systems have made it possible to deal with larger, more complex business situations. Much of this progress has been through the classical techniques of data processing, involving systems analysis and design principles. Many operational systems have of necessity incorporated electronic data processing into their day-to-day workings.

In the preceding several years, the use of computers and word processors for the general office environment has helped users cut costs and increase effectiveness outside the traditional data processing milieu. The authors have been in the forefront in making businesses aware of the benefits of office automation.

As always, new ideas bring new opportunities, and these opportunities should be appreciated by those who must plan, organize, and control the design and implementation of office automation systems. The lessons of the data processing and management information systems communities are only partially applicable to office systems. There are major differences in areas such as end-user interfaces, scope of applications, and development strategies.

We have observed the authors as they have pioneered in the use of sound organizational survey methods to gain real knowledge of how people in their companies were spending their time. We have seen some of their choices of which applications to push forward, and how to best integrate office automation plans with overall corporate goals. These are difficult tasks, and the benefit of the experience of these authors, as related in this book, will help countless others.

The possibilities for office automation are exciting. They range from work-at-home pilot programs (which we and the authors have both found an excellent way of leveraging our workaholic tendencies), to interconnection of all employees via videoconferencing; from the secretary's word processor to the board chairman's wall-size color graphics display; from the creation of an idea to its speedy transmission to all parts of the world. These have been made far more practical as the real prices of microprocessors, memory, and storage devices drop each year. They have been pushed forward through the telecommunications revolution (satellites, cable, digital switching). And they have been fostered by a healthy investment climate.

But what most organizations have lacked is any cohesive planning for the entire range of office automation applications and services. Organizational difficulties have stunted the growth of this field, through such notions as "take all the secretaries away from the principals." And interfacing the increasingly powerful word processing terminals—which provide each end user with more power than most data processing systems had available ten years ago and at extremely low cost—with the corporate information files on the large centralized resources is still a complex problem.

We are glad that the authors have taken the opportunity to share some of their accumulated wisdom with others. Such wisdom is not easy to come by today, and the readers who learn from the lessons and cases in this book will be in a more effective position to assist their organizations in planning and implementing office automation systems. We believe that this exciting field has far to go. The applications in decision support, executive communications, and personal computing we have been working on will build on the methods described by the authors and will will require the methods of planning and organizing that the authors describe. Now is the time to learn.

As Santayana has said, "Those who cannot learn from the mistakes of history are doomed to repeat them." This book offers an opportunity to learn. Take advantage of it.

HOWARD MORGAN
AMY D. WOHL

Preface

For the last several years, the issues and questions relating to the subject of the automated office (office systems) have presented professionals, managers, and executives with a bewildering array of choices and decisions to be made. The demand for a comprehensive and practical guide on the subject of office automation has been expressed by many in the private, public, and academic sectors.

Office Automation has been written by practitioners so that both technical and nontechnical readers can readily understand the opportunities for office automation. Our intended readers include:

Executives wishing a general overview of office automation.

Managers of office automation, data processing, telecommunications, administrative or general services, office services, word processing, records management, and others who need a guide for planning, organization, and control.

Educators, trainers, and researchers responsible for the education and training activities in organizations and institutions.

Programmers, systems analysts, and other professionals involved in all aspects of information systems and resource management activities.

Consultants, vendors, and others involved in the direction and development of office automation products and services.

Office Automation provides the information necessary to make the relevant and conscious choices and decisions based on a systematic and logical sequence of activities that closely parallel the management processes of planning, organizing, directing, and controlling.

This book is the result of years of practical and real-world experiences dealing with all the major problems associated with office automation, information systems, and telecommunications. It is a how-to book that lays out specifics for all phases of office automation. Some of the issues and topics covered are:

Starting an office automation department.

Planning for office automation.

Analyzing the costs, benefits, and opportunities of office automation.

Organizing the office automation activity.

Improving productivity through automation and systematic analysis.

Describing and analyzing the technologies.

Dealing with human factors in automation.

Selling the concept to senior management.

Implementing systems.

Controlling and monitoring office automation activities.

This book is designed as a general reference guide. The guidelines presented are useful in a wide array of industries and organizations. The reader, however, is cautioned to weigh the contents in terms of his or her own environment. The office automation industry is fast paced, with new developments and new products (as well as changes in existing products) constantly coming on the scene. The book has been written so that readers can easily incorporate these new products and developments within this framework. It contains a wide array of checklists, questionnaires, and various other tools to help the user move from concept to implementation.

The views expressed in this book are solely our own and are not necessarily supported, endorsed, or approved by the organizations by whom we are employed.

We should like to offer a collective thank you to the many business, information systems, administrative service and telecommunicatons executives, managers, and professionals who directly contributed to the ideas, concepts, and approaches outlined in the book. In particular, we should like to thank the members of the Office Automation Roundtable for sharing their thoughts with us over the years.

Finally, we are deeply indebted to our families—Phyllis, Camy, Dan, Gabrial, Nancy, Tracy, Michael, Ronnie, Jennifer, and Alison for their inspiration, sacrifice, love, and encouragement throughout this undertaking.

MARK A. LIEBERMAN

GAD J. SELIG

JOHN J. WALSH

New York, New York
Fairfield, Connecticut
New York, New York
March 1982

Contents

APPENDIXES

Part One

INTRODUCTION AND OVERVIEW

CHAPTER ONE

What Is Office Automation?

American business today faces serious issues of productivity, double-digit inflation, and rapidly rising administrative costs. Because of double-digit inflation, the $500 billion in salaries paid annually to white collar and professional employees and the $400 billion paid to clerical and other operational support services will more than double by the end of the decade—to between $15 and $20 trillion.[1]

The following statistics reflect some of the trends of the existing office environment:[2]

1. Administrative costs are climbing at a rate of 10% to 15% a year. Almost 58% of the nation's office expenses are attributable to managerial and professional salaries and fringe benefits.

2. The white collar labor force—one-half the total U.S. labor force of 95 million—is annually growing 20% faster than the total labor force.

3. The demand for information is steadily increasing; each year more than 100 billion telephone calls are initiated and more than 70 billion documents are created.

4. During the past 15 years, productivity, measured by gross national product (GNP) divided by the number of civilian workers, reflected a significant drop, compared to the prior 15 years (15% versus 44%).

The *New York Times* has summarized the concerns about office productivity:

> In 1980, a startling thing happened. For the first time not only did office productivity increases fail to match those of previous years, the output per worker actually declined. The implications are stark. More and more people are employed to generate less and less of the revenue of U.S. business.[3]

Its labor-intensive nature makes the office particularly sensitive to escalating costs. Recent technological developments offer dramatic opportunities to avoid or at least to reduce these escalating costs and to provide information more quickly and more accurately.

4

A recent study by Booz Allen & Hamilton, Inc., suggested that the appropriate application of office automation technology could effect a 15% saving in the time of managers and professionals during the 1980s. Translated into salary-dollars on a national basis, the savings could amount to $125 billion per year by 1985.[4]

Computer and telecommunications technologies have already had a marked effect on the way business is conducted and on the labor requirements in these areas. Over the past 20 years, computer and telecommunications technologies have been successfully introduced into all major corporations in the manufacturing, financial, marketing, planning, and human resource areas. Only within the past five years, however, has industry witnessed the introduction of these technologies into the office environment. Even with these advances, much of the office remains untouched by office automation technology. Consequently, office productivity has stagnated as office costs have continued to increase. For many organizations, these costs represent 50% of all expenses.

Some major reasons for the absence of widespread automation in the office are complex cost justification, office politics, little or no motivation, absence of easy-to-use and "friendly" technology, and difficulty in quantifying and analyzing office tasks and processes. Recent developments and trends in the computer and communications technologies, however, have forced a reexamination of this position.

CONCEPT OF OFFICE AUTOMATION

Although the office and the factory have many similarities, there is one fundamental and very significant difference between them. Generally, in the factory a variety of materials and parts flow on a regular and predetermined schedule through a series of structured processing, finishing, assembling, and quality assurance steps. In other words, the factory represents an integrated environment. In the office of today there is no structure—no clear and easy way of measuring relationships between input and output. The office has developed with little concern for structure and discipline; therefore it has not evolved in an integrated fashion.

The concept of office automation is broad and multifaceted. It consists of multiple technologies (data, text, voice, and image) supporting a broad spectrum of applications (e.g., data processing, word processing, telecommunications networks) that can augment human mental and physical processes.

From a planning perspective, these technologies and applications provide a foundation for integration in the office. Specific attention must focus on analyzing, structuring, blending, and assigning priorities to the human, organizational, economic, procedural, and environmental factors and resources with the available technologies. Such attention is required to obtain maximum benefit without creating confusion and trauma in the work place. Therefore, the change process must be introduced and structured to overcome the resistance to change that is characteristic of people and organizations. Figure 1-1 illustrates this office automation concept.

The goal of integration has not been attained for the following reasons:

1. Incompatibility between different vendors.
2. High costs.
3. Need and justification for automation.
4. Potential violation of privacy.
5. Poor security safeguards.
6. User resistance.
7. Lack of easy-to-use and available software.

In future decades, critical prerequisites for adapting the concept of integration to the introduction and management of change will be an understanding of the complex user environment and the establishment of an appropriate planning and coordination framework. Although variations of the office automation concept may be customized to particular needs, a combination of factors, such as management philosophy, type of industry or organization, product orienta-

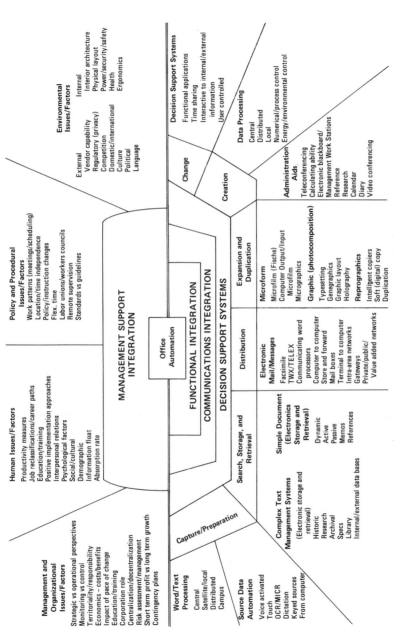

Figure 1-1 Office automation intergration factors.

Management and
Organizational
Issues/Factors

Strategic vs operational perspectives
Monitoring vs control
Territoriality/responsibility
Economics – costs/benefits
Impact of pace of change
Education/training
Corporation role
Centralization/decentralization
Risk assessment/management
Short term profit vs long term growth
Contingency plans

Human Issues/Factors

Productivity measures
Job reclassifications/career paths
Education/training
Positive implementation approaches
Interpersonal relations
Psychological factors
Social/cultural
Demographic
Information float
Absorption rate

Policy and Procedural
Issues/Factors

Work patterns (meetings/scheduling)
Location/time independence
Policy/instruction changes
Flex. time
Labor unions/workers councils
Remote supervision
Standards vs guidelines

Environmental
Issues/Factors

External   Internal
Vendor capability   Interior architecture
Regulatory (privacy) Physical layout
Competition   Power/security/safety
Domestic/international Health
Culture   Ergonomics
Political
Language

Decision Support Systems

Functional applications
Time sharing
Interactive to internal/external
information
User controlled

MANAGEMENT SUPPORT
INTEGRATION

Office
Automation

FUNCTIONAL INTEGRATION

COMMUNICATIONS INTEGRATION

DECISION SUPPORT SYSTEMS

Change

Creation

Data Processing

Central
Distributed
Local
Numerical/process control
Energy/environmental control

Administration
Aids

Teleconferencing
Calculating ability
Electronic blackboard/
Management Work Stations
Reference
Research
Calendar
Diary
Video conferencing

Word/Text
Processing

Central
Satellite/local
Distributed
Campus

Capture/Preparation

Source Data
Automation

Voice activated
Touch
OCR/MICR
Dictation
Keyed sources
From computer

Complex Text
Management Systems

(Electronic storage and
retrieval)
Historic
Research
Archival
Specs
Library
Internal/external data bases

Search, Storage, and
Retrieval

Simple Document
(Electronics
Storage and
Retrieval)
Dynamic
Active
Passive
Memos
References

Distribution

Electronic
Mail/Messages
Facsimile
TWX/TELEX
Communicating word
processors
Computer to computer
Store and forward
Mail boxes
Terminal to computer
Intra-area networks
Gateways
Private/public/
Value added networks

Expansion and
Duplication

Microform
Microfilm (Fische)
Computer Output/Input
Microfilm
Micrographics

Graphic (photocomposition)
Typesetting
Genographics
Graphic layout
Holography

Reprographics
Intelligent copiers
Soft (digital) copy
Duplication

7

tion, economics, level of technical maturity, and people, must be considered in order to determine the nature and the level of automation appropriate for a particular office environment.

ELEMENTS OF THE OFFICE ENVIRONMENT

The office may be defined as the interaction of people to perform a series of processes that require the handling of information with the aid of technology to make business decisions. More specifically, the office is comprised of a series of elements—people, processes, information, technology, planning, directing, organizing, coordinating, and controlling. These elements are grouped and described in Table 1-1.

OBJECTIVES OF OFFICE AUTOMATION

The objective of office automation is to improve the productivity (both the effectiveness and the efficiency) of the office and administrative environment in some of the following ways:

Increase profitability.

Reduce and/or avoid expenses.

Generate time savings.

Attain a competitive advantage.

Support the business environment and organizational goals and objectives more effectively.

Enhance the quality of the work environment of various levels of employees.

Expand the span and control of management.

Provide more tools for better and more timely analysis and synthesis.

Establish a foundation for more effective information integration.

Scope

Peter Drucker has stated that the management of a business corporation must translate its knowledge into effective performance in

Table 1-1 Elements of the Office and Administrative Environment

Elements	Description
People	Levels and classification of personnel involved: managers (all levels), professionals, technicians (research assistants), secretaries, and clerks involved in the processes.
Processes (see Figure 1-1)	1. *Creation*—act of thinking and formulating a communication. 2. *Capture*—placing the communication on a medium (paper, terminal, tape recording) or conveying the message (via a telephone, at a conference). 3. *Preparation*—entry to, processing by, output from, a keyboard. 4. *Revision*—act of changing. 5. *Distribution*—message carrying, mail handling, electronic transmissions, traveling. 6. *Expansion*—copying (reprographics), printing, micrographics (microfilming). 7. *Search, storage, and retrieval*—indexing, storing, searching, information. 8. *Disposal*—discarding and purging of information.
Information	Encompasses the media and form of the information: voice, graphics (charts), images (pictures), data (numbers), text (narrative).
Technology	Involves the electronic, printing, and photographic technologies utilized in an office environment: typewriters, word processors, copiers, terminals, voice automation, computers, communications networks, printers, dictation equipment, photocomposition components, computer output and input microfilm, fascimile, offset printers, graphics,

9

Table 1-1 *(Continued)*

Elements	Description
	optical character recognition (OCR), and others.
Plan/Direct/Organize	The basic managerial functions involved in establishing goals, objectives, and strategies.
Coordinate/Control	Components of the basic managerial functions involved in the coordination of resources to ensure that objectives and action programs are achieved within certain guidelines.

three distinct yet interrelated areas: economic results, productivity and worker achievement, and enhancement of the quality of the environment.[5]

Office automation affects all the areas Drucker suggests. The dimensions of office automation are pervasive in the corporate structure at functional levels (e.g., finance, legal, planning, purchasing, marketing), at user levels (e.g., corporate officer, divisional officer, professional, secretary, clerk), and at organizational levels (e.g., corporate headquarters, division, manufacturing facilities, distribution center). Figure 1-2 illustrates the broad penetration potential of office automation within a typical corporate structure.

Technical Perspective

Office systems are composed of many different technologies, each of which may require the effective use of unique personnel skills. Figure 1-3 illustrates the technical components of the integrated electronic office.

Even though the figure illustrates the importance of communications in linking the many components of office systems, there has been little progress toward achieving their integration. Where integration efforts have been successful, they were preceded by extensive planning activities that established the foundation for the inte-

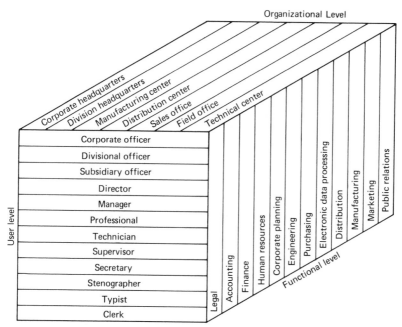

Figure 1-2 Office automation dimensions and corporate structures.

gration. Of course, not all office automation activities may require integration. Many technologies may stand alone in support of local or departmental requirements; others may require only partial integration, such as a word processor to a photocomposition device.

Integration does not require that all components reside in one physical location. This characteristic will provide an organization with the flexibility to accomplish certain office activities independent of a fixed or centralized work environment and thus to optimize the use of the demographically changing work force and lifestyles of the population. Tomorrow's offices will offer a choice of the workplace—for example, the corporate office, the home, the airplane, the hotel, the suburban office. On the one hand, mobility will bring about greater flexibility; on other hand it will create new management problems, for example, control of "delivera-

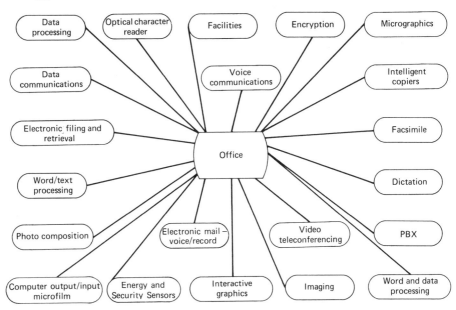

Figure 1-3 Technical components of office automation.

bles" when a manager and a subordinate are not in the same location.

USERS OF OFFICE AUTOMATION

Computerization in the office can be viewed from a number of perspectives.[6] It varies by job level, job function, degree of support required, and other factors. Four classifications of potential office systems users are defined.

Executive Management

As a key decision maker in the organization, the executive needs timely and accurate information to meet business objectives. The day-to-day functions and activities of the executive are relatively unstructured, unpredictable, and not very repetitive. Much of the executive's time is spent attending meetings, absorbing information, negotiating, talking on the telephone, solving difficult problems, and making decisions.[7]

The attempts of the data processing community in the mid-1960s and early 1970s to provide management information systems directly to executives were largely unsuccessful, because the systems and technology were complex and were not designed for executive end-user utilization and operation. The initial benefit of office automation to the executive will be realized indirectly through the increased effectiveness and productivity of middle and first-line management and professional personnel, because of their access to more accurate and more timely information. In time, with the establishment of better and simpler tools and with better education, the direct use of automation by senior management may become a reality.

Middle and First-Line Management

Historically, data processing has been more supportive of middle and first-line managers than of the executive. Much of the current electronic data processing (EDP) activities has been directed toward providing middle and first-line managers with information helpful in controlling and directing the day-to-day operations of the organization. Generally, application penetration has been predominant in accounting, inventory, personnel, marketing, and manufacturing.

More support is needed for timely managerial information: better integration of textual and numeric data into a base of information; improved communications with peers, subordinates, and superiors; and better access to existing information sources, both inside and outside the organization.

Middle and first-line managers are also becoming more aware of the computer's potential, envisioning and demanding the kinds of services they believe the computer can supply to their organizations. Demands for quicker and cheaper communications capabilities are increasing, as is penetration of quality color graphic displays, new printing devices, and other graphic media.

Professional and Technical Personnel

A significant amount of the work in any organization is relatively unstructured knowledge work normally performed by professionals. Such work deals with ideas in support of the management functions of planning, policy making, coordination, and control. This work is a prime target for a number of support tools, such as:

1. The use of keyboards and displays to replace pads and pencils in the process of crystallizing ideas into summaries and conclusions.
2. The use of graphics composition tools, displays, and analytical computing aids to model, compare alternatives, understand relationships, and develop presentation materials.
3. The use of electronic media to store, cross-index, retrieve, and dispose of reference materials.
4. The use of electronic communications to share ideas and information and to conduct meetings without the physical presence of the participants.

Secretarial and Clerical Personnel

The word and text processing industry has selected the secretarial/-clerical role as a first candidate for automation—originally in the form of improved text creation and editing and more recently with electronic filing, mail, calendars, scheduling, telephone answering, and intelligent copier products.

Much has been made of the need not only to automate the secretarial function, but also to break it down into typing and administrative components. Two rationales have been presented for this approach: (1) productivity gains to be achieved by making typing a production job, which has been necessary to justify equipment costs and (2) the need for creating secretarial career paths. However, the close association among the secretary, the manager, and the staff has real value and should probably not be tampered with initially—and then only very gradually.

Clerical personnel perform a variety of nontyping functions that can also be improved by modern technology. Whether the clerical work is in the accounting, personnel, legal, manufacturing, marketing, or any number of other departments, technology has been and is being designed to supplement and reduce many of the structured and repetitive clerical activities.

PHASES AND STAGES OF OFFICE AUTOMATION

In researching for more effective ways of integrating and blending new technologies into existing organizations, a lesson can be learned from history, economics, and business cycles.

John Flaherty describes the evolution of a business in three phases: the traditional phase, the transitional phase, and the transformational phase. The intent is to develop a diagnostic tool for introducing and planning change, in order to assure the continued and healthy growth of an organization in a conscious and purposeful way.[8]

The seminal work on attempting to understand the various stages through which organizations have evolved in their use of computer technology was developed by C. F. Gibson and R. L. Nolan. These authors refer to the four stages of evolution and show how a typical company progressed, from the introduction of data processing technology in an initiation stage through a mature development stage.[9]

In a more recent article, Nolan expanded the original four stages to six stages, in an attempt to explain the continual rise of EDP expense levels and new technological innovations.[10] Others have also attempted to explain the evolution of office automation in different ways.[11]

To illustrate the characteristics of the evolution of office automation, the authors have used a framework consisting of three broad phases—the traditional, the transitional, and the transformational—each of which consists of one or more development stages. Table 1-2 illustrates the phases and stages of office automation and their respective characteristics.

Table 1-2 Phases and Stages of Office Automation Evolution Characteristics

Phases	Stages	User and Management Attitude, Involvement and Commitment	Technology Usage	Planning and Control System in Place	Responsibility, Organization, Skill Requirements and Management Characteristics	Application Scenarios
I. Traditional	1. Initiation	Hands off attitude Little support Disinterest	Uncoordinated Multiple vendors Technology used as toys Incompatible technologies	Lax management Unclear jurisdiction Routine operational support	Fragmented, non-macro perspective with specialized skills Tinkerer	Obvious cost reductions Applications (word processing/typing pools) Isolated applications
II. Transitional	2. Expansion	Participants but dependent on technical resources Growing expenses Focus on how people do things Better administrative support	Simple non-complex usage Fragmented and uncoordinated activities	More lax management Short-term budgetary controls Start of education	Specialization to develop a variety of applications Power struggle between EDP, office administration and others	Uncoordinated proliferation in all areas (word processing/electronic mail) Electronic storage and retrieval Task orientation versus semi-structured automation

16

	3. Formaliza-tion	Cooperative participation/ sponsorship Focus changes from how people do things to what they do (processes)	Functional work stations (legal, accounting, manufac-turing, etc.) Simple electronic file access	Short-term planning and control oriented guidelines Explosion of education	Assignment of office automation responsibil-ities and development of charter Technician	Knowledge worker augmentation Distribution of administra-tive zones Opportunity displacement Structured and semi-structured automation
III. Transforma-tional	4. Integration	Functional specializa-tions Full involvement and sponsorship	Cost effective voice, touch, other source data entry and conversion to digital media	Start of long-term and strategic planning and control orientation Project management Development of standards	Generalists and specialists evolving Conceptualist and start of credibility Full organization status	Start of integration applications, systems, w.p., reprographics, mgt. work stations, digital networks with EDP

18

Table 1-2 *(Continued)*

Phases	Stages	User and Management Attitude, Involvement and Commitment	Technology Usage	Planning and Control System in Place	Responsibility, Organization, Skill Requirements and Management Characteristics	Application Scenarios
	5. Maturity	Take lead in design activity Distributed responsibility	Functional integration of office/EDP/ telecommunications applications Semi-structured/ unstructured automation	Macro and micro level integration authority Information resource policy Maintain and expand credibility	Resource and profit orientation Reorient expense to investment decisions for office automation Adopt office automation strategy as major internal business strategy	Major use of semi-automated and automated administrative aids for decision support activities on a global scale

Phase I—Traditional

Phase I (consisting of Stage 1) of office automation represents much of the current state-of-the-art thinking in most organizations. The major focus of this phase is on secretarial and clerical personnel, in fragmented environments and with narrow perspectives. This phase represents the initiation stage.

Stage 1—Initiation

In this stage, organizations perceive technological opportunities for cost reduction and increased productivity through the use of traditional office equipment. This stage generally focuses on word processing systems. Emphasis is placed on the more efficient production of paper (as opposed to the longer range objective of increasing administrative productivity). In most organizations, the technology is introduced and managed by administrative services on a centralized basis. At this point, a fairly clear delineation exists between word processing and data processing.

The duration of phase I may vary from one to three years, depending on the organization, its environment, and its level of technical sophistication. Some users are taking advantage of the word processor's abilities to interface to photocomposition, communications, and other systems. Many of today's large organizations are in this stage. The traditional phase is followed by the transitional phase.

Phase II—Transitional

The transitional phase (Stages 2 and 3) involves taking risks to potentially realize greater opportunities. This phase consists of two stages, expansion and formalization.

Stage 2—Expansion

The initiation stage is followed by a stage of accelerated office systems expansion. Office systems will catch on, just as data processing did during the second stage of EDP growth. Interest in and costs of automation will increase. Poorly planned and uncoordinated development efforts will proliferate, unless certain preventive actions are taken.

As office systems increase, organizations are being forced to deal with the formal organizational issues. The responsibility struggle for the office systems function between various functional groups is heightened as management recognizes the potential opportunities. The duration of this stage may also range over a period of one to four years, depending on the environment.

Stage 3—Formalization

The third stage of the office automation evolution, which is another part of the transitional phase, encompasses more formalized analysis focused on administrative problems and functions. Office systems planning, development, and operational responsibilities will be formally assigned during this stage. As this stage evolves, it is unlikely that full responsibilities for office systems will be assigned to one person or one department. However, as office systems expenses increase, this responsibility will change over time to a more directed focus.

A major thrust during this stage should be on education. Additional efforts should be devoted to developing the appropriate analytical tools and methodologies to more accurately analyze the work flow patterns, expense streams, and other office components.

In this stage, applications should emphasize improvement in knowledge worker productivity. Distributed administrative support should become commonplace. Electronic filing, storage, and communications capabilities will proliferate rapidly.

The completion of the transitional phase should lead to the start of the transformational phase.

Phase III—Transformational

During the transformational phase (Stages 4 and 5), office systems should have penetrated most organization levels and its application should have been highly integrated. The transformational phase consists of the integration and maturity stages.

Stage 4—Integration

Stage 4 requires management that can plan, implement, and maintain integrated office systems. Although not all areas require integration, it must nevertheless be a prerequisite for this stage.

Because the expenditure levels and the impact of office automation will be so significant, the integration stage goes beyond most current management perception. Therefore, formal planning, control and operational responsibilities for office systems should be assigned by management to establish a cohesive direction.

In this stage, software, hardware, and communications will be available on an organization-wide basis. To support integration, various types of interfaces between word processors, data processors, photocomposition, intelligent copiers, electronic files, and external data bases should be commonplace and easy to implement.

The use of multifunctional work stations (e.g., manager, secretary) should have significantly penetrated the secretarial, clerical, and professional levels. Some work station penetration of middle- and executive-management levels will also have occurred.

State 5—Maturity

The final stage in the evolution of office automation is maturity. At this stage, the concept of corporate information resource management (IRM) involving a convergence of office automation with telecommunications and information systems and services should become a reality and not merely a concept. Companies may consider adopting this area as a major internal business strategy and start to evaluate this investment and philosophy, in competition with all other business investments.

At this time, it is not certain how fast or how far this convergence and integration will progress, but certain specific actions can and should be pursued to take advantage of new opportunities, to better manage the coming and inevitable changes, to position an organization to minimize risks and uncertainty, and to overcome the problems, obstacles, and constraints that have historically plagued the data processing industry.

CONSTRAINTS AND OTHER ISSUES

Significant nontechnical issues must be addressed if office systems are to effectively penetrate organizations. Ergonomic, privacy, behavioral, career, union, societal, safety, demographic, security, educational, and health factors and issues may have constraining influences on the growth and success of office systems if these matters are not adequately resolved. Some examples follow:

1. Office systems implementation will lead to a significant reexamination of existing educational efforts, corporate training programs, work habits, and other policies and procedures.

2. Business conditions and office technologies will accelerate trends toward decentralized support of business operations. This will require a critical examination of the centralization and decentralization issues (see Chapter 5).

3. Increased attention must be given to facilities planning, aesthetics, space, and the appropriate levels of heat, noise, temperature, humidity, and color to assure human comfort. Otherwise, the technology may result in discomfort and work disruption (see Chapter 8).

4. Health and safety issues must focus on postural, visual, audio, and other human comfort factors (see Chapter 10).

5. For those employees who are displaced by office systems, organizations should sponsor educational and retraining programs as part of their social responsibility (see Chapter 10).

6. To reduce resistance to change, positive implementation strategies must be developed through better communications and participation. Fear of the unknown and of job security must be addressed positively by executive management.

KEY IDEAS

This chapter has provided an introduction to office automation. Accelerating office costs, increases in head count, and productivity opportunities coupled with advances in technology are the primary motivators for introducing office automation.

Office automation covers a broad spectrum of technologies and raises many organizational, economic, procedural, and other issues. It is multidimensional and will evolve through several phases and stages, each with its unique characteristics and time dimensions. In each phase, management, technical, economic, and behavioral issues must be overcome to reach the next phase. No phase or stage can be avoided. However, those responsible for office automation must be prepared to consciously manage each phase and stage to minimize disruptions, reduce risks, and focus in an appropriate direction.

The remaining sections of the book are organized in a logical and easy-to-use manner. Part II contains information on how to plan, organize, coordinate, and control office automation activities, as well as how to identify opportunities, benefits, and costs. The topics of needs analyses, requirements definition, and implementation strategies and tools are presented in Part III. Finally, Part IV reviews office systems technologies and discusses the future implications of office automation.

REFERENCES

1. Gerald Tellefson, "Productivity Revisited," in *Summary of Proceedings*, Office Technology Research Group, Pasadena, CA, September 1980.

2. International Data Corporation Special Report, "Productivity and Information Systems for Tomorrow's Office," *Fortune*, September 1980.

3. Randy J. Goldfield, "The Office Today: The Drive for Productivity," *New York Times* Advertising Supplement, June 28, 1981, p. 6.

4. Tellefson, *op. cit.*

5. Peter F. Drucker, *Management—Tasks, Responsibilities, Practices* (New York: Harper & Row, 1974).

6. D. Drageset, "Users of Office Automation," Presentation at AIIE Office Automation Conference, New York, 1978.

7. Henry Mintzberg, *The Nature of Managerial Work* (New York: Harper & Row, 1973).

8. John E. Flaherty, *Managing Change* (New York: Heller, 1979).

9. C. F. Gibson and R. L. Nolan, "Managing the Four Stages of EDP Growth," *Harvard Business Review*, January—February 1974.

10. R. L. Nolan, "Managing the Crises in Data Processing," *Harvard Business Review*, March—April 1979, pp. 115–126.

11. See, for example, P. A. Strassman, "Stage of Growth," *Datamation*, October 1976; Michael D. Zisman, "Office Automation: Revolution or Evolution," *Sloan Management Review*, Spring 1978, Vol. 19, No. 3, 1–16; F. W. Holmes, "IRM–Organizing for the Office of the Future," *Journal of Systems Management*, January 1979, pp. 24–31; Christopher J. Burns, "The Evolution of Office Information Systems," *Datamation*, April 1977; Robert I. Baxter and George F. Krall, "Six Stages Toward the Automated Office," *Words*, October–November 1979, pp. 20–24; and James H. Bair, "Communications in the Office of the Future: Where the Real Payoff May Be," *Business Communications Review*, January-February 1979; Vol. 9, No. 1.

Part Two

PLANNING, ORGANIZING, AND CONTROL

CHAPTER TWO

How to Begin and Plan?

Traditional office processing methods are no longer cost-effective for today's business needs. Careful planning is necessary to ensure the proper initiation and continuation of an office automation program.

The initiative to perform some preliminary research in the area will generally come from management. To begin the program, the initiator is faced with a bewildering array of options, alternatives, priorities, and constraints.

The first steps in starting an office automation effort are to establish a reference base, to develop an understanding of the concepts, to establish preliminary objectives, and to develop a systematic approach to achieve the objectives.

A general framework for office systems planning must address the following questions:

Where are we? This step provides a reference base of the external and internal pressures on the office. It also provides a profile of the business and the existing internal administrative office in terms of costs, resource levels, equipment, inventories, strengths, limitations, organization structures and relationships, needs assessments, and management philosophies.

Why change? After a thorough analysis and understanding of the reference base, the pressures for change that are affecting traditional thinking must be evaluated. This step identifies and addresses major objectives, issues, and office automation opportunities.

What can we do? This step requires an evaluation of both strategic and tactical alternatives in terms of organization structures, missions, applications, priorities, resource requirements, benefits, risks, constraints, and key assumptions.

How do we get there? This step provides the recommended action programs and selected alternatives, along with strategies, resource allocations, establishment of priorities, action programs, and monitoring and control mechanisms.

Figure 2-1 illustrates the flow and components of the office automation planning framework. Figure 2-2 represents a convenient, clear,

28

and short way of presenting the results of the planning process to management.

ESTABLISH A REFERENCE BASE

The following techniques should be used by the reader to become acquainted with the basics of office automation:

1. Read current periodicals, books, special reports, and subscription service literature (e.g., *The Office, Administrative Management, Business Week, Datamation, Fortune, Forbes, Words, Data Communications, Infosystems, Computerworld, Seybold Reports, Data Quest, Diebold Group, Electronic Mail and Message Systems, Advanced Office Concepts Newsletter*).

2. Attend schools offering courses in office automation and seminars (e.g., American Management Association, Datapro, Auerbach, American Institute of Industrial Engineers, National Micrographics Association, National Computer Conference's— Office Automation Conference, Massachusetts Institute of Technology, Harvard University, Institute for Graphic Communications).

3. Join one or more professional organizations that focus on some or all aspects of office systems (e.g., American Federation of Information Processing Societies (AFIPS), International Information Word Processing Association (IIWPA), International Communications Association (ICA), Data Processing Management Association (DPMA), Society for Management Information Systems (SMIS), Office Automation Roundtable, Diebold Group, The Gartner Group, Office Technology Research Group). Before joining a professional organization, make sure it has a charter and that it is related to the organization's objectives.

4. Join or subscribe to one or more office automation research groups. Review the services of the Diebold Office Automation Group, International Data Corporation, Profit Oriented Systems Planning Group, and Quantum Science, and consulting firms such as Stanford Research Institute, A. D. Little, Arthur

30

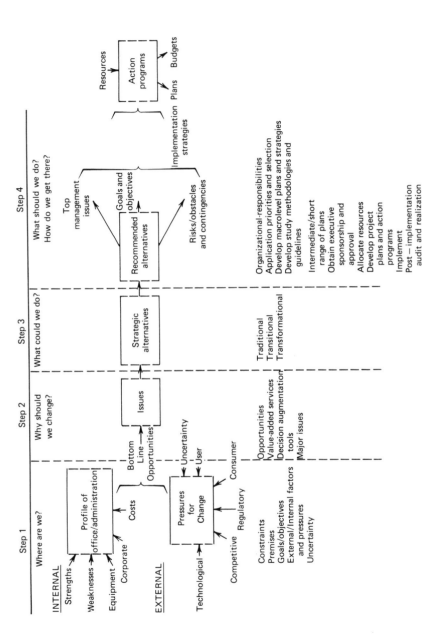

Figure 2-1 Office systems planning framework diagram.

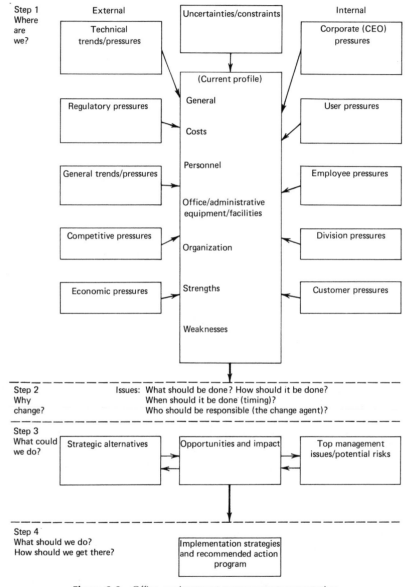

Figure 2-2 Office environment management presentation.

Anderson, and the Yankee Group. Some organizations are conceptual; others are pragmatic. Some provide publications as well as workshops.

5. Attend conferences where promotional material is displayed—for example, IIWPA's Symposium, National Computer Conference-Office Automation Conference, International Communications Association Conference, American Institute of Industrial Engineers Conference, Clapp & Polliak's Annual Infosystems Conference,

6. Contact office automation consultants to develop a firsthand knowledge of office systems. It may also be necessary to hire a consultant to gain management's attention, to present an external perspective, or to acquire a special talent for performing a specific assignment.

7. Meet with vendors whose products and services include computers, communications, micrographics, reprographics, electronic storage, software, and related products and services. Talk to vendors' marketing, product-planning, engineering, and executive personnel.

8. Meet representatives of other companies and other organizations to assess where they are and what they have done. Consider arranging small informal intercompany office systems roundtables to share experiences.

Certain basic information should be accumulated in order to establish a reference base, develop plans, and obtain management approval and commitment for a sustained office automation activity.

External Factors, Pressures, and Trends

Several environmental trends and pressures affect office automation. Each trend or pressure may not impact all organizations in the same way because of differences in products or services, type of industry, degree of labor or document intensity, size, geographic dispersion, degree of vertical or horizontal integration, customer composition, and other reasons. A knowledge and awareness of such external pressures and trends as technology, demographics, economics, social factors, competition, and regulations must be accumulated.

Internal Factors, Pressures, and Trends

Although internal pressures may not always lead to an effort to initiate office automation, they certainly can help and are often essential. Internal pressures and trends can emanate from several sources, such as the executive office, department managers, secretaries, professionals, and others.

Examples of some internal pressures and trends include the following:

1. Improve administrative productivity and reduce costs.
2. Address the critical shortage of support personnel.
3. Reduce the time to move information from sender to decision maker (information float).
4. Improve customer service.
5. Create a better working and career path environment.

Understand the Organization

Several key organizational ingredients for successful office automation are:

1. Understand the power and influence bases.
2. Identify the key formal decision makers.
3. Identify management, professional and secretarial personnel who influence the key decision makers.

Because of these factors, it is important to identify all the key personnel (information systems, telecommunications, human resources, and others) so that they can be informed of, and invited to participate in, the initial effort, as appropriate. This is particularly important in multidivisional and multinational organizations, so that "everyone who counts" participates in the planning process.

Consider Constraints, Risks, and Uncertainties

It is important that the plan include a section on major office automation constraints, risks, and uncertainties, such as the following:

Constraints

Current equipment, personnel, systems, and service base.

Resistance to change (e.g., manager/secretarial relationships, unfriendly human/machine interfaces).

Absence of technological integration because of absence of standards.

No meaningful measurement of office productivity.

Limited office expense tracking mechanisms.

Absence of significant office automation successes.

Lack of formal office systems responsibilities.

No management awareness.

No strategic focus.

Risks and Uncertainties

Risks exist at all levels. There are risks in doing something or in not doing something, and these can impact the bottom line. The planner must be aware of these risks and attempt to minimize them. Without a knowledge of these risks, the planning effort will not be complete. Here are a few examples:

Benefits may not materialize and the costs of automation may be underestimated.

Selection of inappropriate vendors may reduce the integration potential and result in greater incompatibility.

Establishing of security and backup measures for all aspects of office automation may be difficult and expensive.

Designing of fail-safe systems may also be difficult and expensive. Equipment and power failures may cause business disruptions.

Privacy of data may be violated because of the distributed and decentralized nature of the office.

Risk of nonaction may result in the loss of a competitive advantage.

Recognize Strengths and Limitations

It is important to assess the strengths and limitations of the existing office environment. Strengths may be leveraged and used to gain significant advantages; limitations must be overcome. Some examples of strengths and limitations are:

1. Degree of business maturity and level of sophistication of the data processing, telecommunications, and administrative functions.
2. Corporate profitability and controls.
3. Availability of capital and human resources.
4. Absence of formal responsibility and authority for office automation activities.
5. Absence of strategic and tactical office systems plans.
6. Existence or nonexistence of corporate standards.

As a result of the reference base analysis, the following actions should be taken:

1. Develop the preliminary office automation program objectives and recommendations.
2. Establish a plan for achieving the objectives.
3. Present the plan and action program to management for approval.

ESTABLISH PRELIMINARY OBJECTIVES

Specific objectives as enumerated below can be established (1) by an individual who recognizes the opportunities of office automation, (2) by an informal group with a common interest, (3) by a formal multidisciplinary steering committee, or (4) as a result of endorsement by senior management.

1. Establish the scope of the initial effort. For example, will the initial effort be company wide or limited to a division or specific function?

2. Identify pilot application areas (e.g., word processing in personnel, electronic mail in marketing).

3. Obtain management approval and sponsorship.

4. Who has responsibility and control? How do systems, telecommunications, administration, and other departments interface —formally and/or informally?

5. How is the organization structured (centralization, decentralization, matrix)?

6. What are the administrative expenses and where are they concentrated?

7. What are the past, current, and projected head counts and salaries by job classification, functions, and locations?

8. What are the major administrative processes, information flows, and issues?

9. Establish a preliminary office systems plan to guide short-term decisions on pilot selection, vendor selection, and application selection and priorities.

10. Evaluate office systems vendor products, services, and consultants.

11. Develop an education program.

12. Develop an internal promotion campaign to familiarize and orient management and nonmanagement personnel.

DEVELOP A PLAN

Planning encompasses several time periods. Strategic planning covers a long time frame; tactical planning covers an intermediate time frame; operational planning covers a short time frame. Planning involves examining issues, setting objectives, and developing plans necessary to achieve the objectives. (What is going to be done? Who will do it? How will it be done? With what resources? Within what time period?)

Some objectives of office systems planning are:

1. To support the business environment and organizational goals and objectives.
2. To communicate the concept of office systems from management, technical, and administrative perspectives.
3. To identify key issues and resolve them.
4. To identify savings opportunities and realize them.
5. To establish a strategic direction and framework for the automated office for the next three to seven years.
6. To establish a series of strategies to maximize the broad goals of increasing profits, increasing productivity, reducing information float, reducing/avoiding costs, and reducing risks.
7. To better anticipate and plan for uncertainty and change.
8. To integrate the functional areas of information systems, telecommunications, and office technology within the organization.

Office automation must be approached from a strategic basis in order to define implementation phases that identify an overall integration direction, specific projects, and strategies, as well as resource requirement and benefits. At a minimum, any office systems strategy must focus on funding cost-justified projects, recommending organizational approaches, and obtaining management commitment along the way. Table 2-1 illustrates five broad office automation planning phases and their technological components. As shown, these phases can overlap. The rate of development and evolution will largely be a function of resource availability and the absorption capacity of the organization to accommodate change. Some other major office systems planning considerations include the following:

Table 2-1 Office Automation Planning Phases and Technology Readiness

	Year	1	2	3	4	5	6	7	8
1. *Initiation Phase*		▮	▮						
Form steering committee									
Exploration/feasibility reports									
Pilot/assessment projects									
Administration/office strategy/plan									
Information resource management issues/policies									
Education/training—all levels									
Stand-alone/shared word/text processing									
2. *Expansion Phase*					▮	▮			
Central/distributed word and text processing									
Electronic mail networks									
Computer access—internal/external									
Administrative zones									
Photocomposition/OCR interfaces									
Electronic filing									
3. *Formalization Phase*						▮	▮		
Administrative/professional work stations									
Word/phrase dictionary									
Calendar management/electronic blackboards									
Intelligent copier access/graphics									
Data-base storage/retrieval/access									
Interactive graphics									

4. *Integration Phase*
   Management/professional multifunction work stations
   Visual teleconferencing/meetings
   Direct data entry (voice/writing/graphics)
5. *Maturity Phase*
   Decision support systems
   Ad hoc information and data-base access
   Home telephone/terminal cable access
   Personal computing integration

1. Establish a full-time multidisciplinary office systems department with a specific charter and mission (as discussed in Chapter 4).

2. Form an office automation advisory (steering) council to identify specific problems, strategies, and implementation solutions, to guide progress, to promote communications among all interested groups, and to approve plans and major projects. This steering council must consist of senior management from each major operating and staff unit. The steering council must represent users (not technologists) and must be able to represent their respective divisions.

3. Develop the appropriate office automation planning and analysis tools (e.g., survey techniques, study questionnaires, study guidelines, measurement, schedules, and programs) (see Chapter 6).

Table 2-2 provides a sample outline of an office automation plan.

KEY IDEAS

To cope with the numerous pressures and factors associated with office automation, specific objectives and a cohesive plan for meeting these objectives are required.

To be successful, the approach must be phased, must be cost-effective, must have management sponsorship, must identify the "deliverables," must be measurable, must support the business goals, and should not dramatically differ from any already established management philosophy. The planner must recognize the constraints, risks, and uncertainties associated with the effort, and must be aware of the characteristics of successful as well as unsuccessful office automation programs.

Several approaches to office automation are available to the planner. One alternative is to take a pioneering or leading-edge approach to effect the change. A second alternative is to take a passive or skeptical approach. This organization will wait and see what is going on at other organizations before making any decisions. Mid-

Table 2-2 Office Automation Plan Outline (sample)

1. Executive summary
2. Introduction
   (a) Purpose and objectives
   (b) Scope and limitations
   (c) Plan methodology and work conduct
   (d) Major issues being addressed
   (e) Major internal and/or external environmental pressures, factors, and trends
   (f) Key assumptions, constraints, and uncertainties
3. Current profile of office automation environment within organization (reference base)[a]
   (a) Major business objectives and strategies
   (b) Overall administrative/office expense profile
   (c) Office automation equipment and facilities profile
   (d) Office automation organizational responsibilities
   (e) Current strengths, limitations, and concerns
   (f) Characteristics of the office automation environment in other organizations
4. Major office automation business/functional support opportunities, strategies, and priorities
   (a) Macroassessment of opportunities by company, subsidiary, division, department, and function
   (b) Macroassessment of costs/benefits
   (c) Selection of priorities—strategic, tactical, and operational programs
   (d) Office automation requisites and strategies
   (e) Plan, budget, and other monitoring linkages
   (f) Organization responsibilities
   (g) Vendor selection—make or buy strategies
5. Project plans and budgets
   (a) Functional/application candidates and milestones
   (b) Division/corporate candidates and milestones
   (c) Operating/facilities/hardware/software plan
   (d) Head count plan
   (e) Education/training plan
6. Financial Plans
   (a) Budget and discounted cash flow analysis
   (b) Resource requirements by type (personnel, facilities, equipment, etc.)
   (c) Projected return on investments, and/or payback calculations by project(s)

41

Table 2-2 *(Continued)*

7. Appendix/references (if required)

[a]An optional section could include a profile of what other organizations are doing or planning to do in office automation.

way between these alternatives are other alternatives and options, which vary depending on the environment, resources, needs, style, and level of technical sophistication of the specific organization. Whatever alternative is ultimately selected, a vital and critical initial step is to develop a plan and start the action programs.

42

CHAPTER THREE

How to Identify Preliminary Costs, Benefits, and Opportunities?

The quantifying and qualifying of potential benefits and cost-saving opportunities through the introduction of office systems technology can best be achieved through an internal assessment program. The results of this assessment will provide the framework in which actual savings can be determined. Much of the research material that has been developed about the office has been too generalized to be considered for inclusion in a benefits analysis for presentation to management. Most of the studies have concentrated on the area of text processing or have been primarily qualitative. Some studies have made detailed analyses of management, professional, clerical, and secretarial activities; however, because of wide variation among the functional needs in different industries, it is difficult to simply extrapolate published statistical results and apply them in any general fashion.

This chapter is mainly intended to provide a framework for identifying areas of cost, benefit, and opportunity. Part III provides a more detailed framework for analyzing, classifying, and matching the specific application needs with available technologies.

It is also necessary to analyze such areas or processes as mail, telephone use, and filing system characteristics. Moreover, it must be noted that it may be a gross error to assume that all organizations have similar or identical needs.

There are several important aspects of conducting an office needs study. The first step is formation of a study team, which should consist of personnel with experience in business systems, planning, statistics, human factors, and similar disciplines. The second is implementation of a variety of available survey tools and methodologies. These are discussed in Chapter 5.

Other aspects include the use of statistically valid measurement techniques, access to data entry and analysis support, and internal promotion of the survey effort to solicit the necessary internal support.

One of the first steps in conducting the study should be the development of a profile of expense categories that can potentially be reduced through implementation of office systems applications. Certain preliminary and basic information must be gathered, accumulated, and analyzed to establish the proper reference base for identifying opportunities, benefits, costs, and potential constraints.

SOURCES AND IDENTIFICATION OF INFORMATION

Only a broad and systematic analysis will enable the office systems planner to identify opportunities over an extended period of time.

Because such an analysis may be costly, time-consuming, and sensitive, senior management must approve it before the analysis is conducted. The data for the study may be gathered from a variety of sources. Among them are:

The annual report of the company

A copy of the Form 10K statement filed by the company

Corporate archives

General corporate information centers

Corporate staff profiles

Operating unit profiles and strategic plans

Corporate personnel data

General ledger financial data

Fixed assets records

Corporate manuals

Organization charts

Management and planning reports

Additional information about management, professional, secretarial, and clerical needs will also have to be developed.

Specific data will have to be gathered about how managers and professionals spend their time—reading, writing, attending meetings, receiving and responding to communications, searching files, planning, and other activities. This information will have to be analyzed by level, by department, by function, and by other categories (see Chapters 6 and 7).

Delegable Activities

An effort should be made to determine the amount of managerial or professional work that could be delegated to the secretarial or the other staff. This result should be correlated against the administrative tasks performed by both categories to determine cost savings opportunities.

Managers and professionals must be questioned carefully and in detail concerning the types of problems they encounter. For instance, in previous studies a major problem category was the large number of interruptions most managers encountered routinely, which hampered their ability to concentrate on regular business matters. Another typical complaint relates to information. This usually translates into a need for improvement in the timeliness and availability, format and accuracy, and retention of information needed to perform managerial or professional functions.

Specific data will also have to be gathered on how secretaries and clerical staff spend their time—typing, filing, copying, scheduling meetings, and performing other duties. These data will have to be further analyzed on the basis of the number of managers and professionals supported. Active procedures for specific tasks (i.e., handling mail, taking messages, etc.) will also have to be studied to provide an appropriate understanding of the levels and types of administrative support services received and/or really required by managers and professionals. Problems and constraints should also be probed so that any inadequacies that exist may be identified and addressed appropriately.

Information categories

It will be necessary to identify the most commonly used processes, procedures, and techniques for the creation, capture, preparation, distribution, storage, copying, and destruction of documents and other data. In addition, basic data on the types of documents and volumes, and their original and ultimate destinations will have to be included. Where appropriate, problems associated with existing procedures should be identified and recommendations for improvements developed.

The following partial list of formal and informal office tasks, processes, and activities clearly reflects the broad scope of individual actions that occur within the office environment and which are candidates for review and improvement:

Addressing
Appraisals
Calculating
Calendar maintenance
Charting
Collating
Data entry
Dictation
Distribution
Distribution list maintenance
Documentation
Editing
Errands
Expense reporting
Flowcharting
Follow-up files
Forms preparation
Graphics
Information inquiries
Indexes
Inventories
Itineraries
Library
Mail logging
Mail handling
Manuals
Meetings
Message preparation

Organization charts
Originating memoranda
Personnel information and procedures
Petty cash coupons
Plans
Presentations
Priorities for work
Procedures maintenance
Programming
Proofreading
Reading
Records management
Report Preparation
Research
Reservations
Schedules
Special communications needs
Stationery/supplies control
Telephone responses/directories
Time card processing
Transportation requests
Training
Travel
Waiting for work

The National Bureau of Standards has established classifications of typical "office products," which are listed in Table 3-1.

Other examples of the work of the National Bureau of Standards are shown in Tables 3-2 and 3-3, which illustrate a framework for identifying and ranking potential productivity improvements based on key product time and cost factors. In these examples the office automation analyst identifies key products and the work activities (e.g., professional time, support time) associated with the preparation process.

Preliminary Information Classification

Once the data have been documented, they must be classified into meaningful categories for analysis and evaluation. These classifications include:

Direct expenses

Indirect expenses (e.g., rent, utilities, mail utilization, supplies, office equipment)

Job classification by level:
  Executive, middle, and first-line management
  Professional and technical
  Secretarial/clerical

Functional categories:
  Legal
  Research and development
  Finance
  Engineering
  Human resources
  Corporate staff
  Marketing

Organizational structures:
  Centralized
  Decentralized
  Matrix

Table 3-1 Typical List of Office Products[1]

Correspondence	Letter
	Memorandum
	Message
Reports	Management
	Trip
	Technical
	Incident
	Project status
	Fiscal
	Personnel
	Weekly activities
	Material deficiency
	Training
Documents	Statement of work
	Specifications
	Procurement plan
	Program management directive
	Program management plan
	Letter request
	Sole source justification
	Determination and finding
	Invitation for bid
	Request for proposal (RFP)
	Equal Employment Opportunity (EEO) certification
	Small business coordination
	Preaward survey
	Model contract
	Change order
	Administrative notice
	Source selection plan
	Annual call for estimates
	Procurement directive
	Delivery order
	Cost estimate
	Independent cost analysis
	Contract funds status report
	Staff meeting agenda (and report)
	Action item list
	Configuration change status report
	Engineering change proposal (ECP)

Table 3-1 *(Continued)*

	Quarterly resources report
	Systems safety program plan
	Configuration control board minutes
	Data management report
	Training plan
	Program management systems checklist
	Contract management systems checklist
	Site survey report
	Environmental assessment
	Life-cycle cost study
	Phase-out plan
Forms	Security classification guide
	Inspection and acceptance document
	Data item description
	Personnel action request
	Time card
	Work order request
	Security monitor
	Printing request
	Position description
	Purchase request
	Report of survey
	Travel request
	Military order
Reviews/Briefings	Business strategy panel meeting
	Quarterly financial review
	Periodic program review
	Project management review (PMR)
	Executive management review (EMR)
	Resources utilization committee action
	Financial management board review
	Division advisory group review
	Internal management review
Audiovisual Aids	Vugraph
	Briefing text
	Graphic Aid
	35mm slide
	Briefing board

Table 3-2 Annual Personnel Costs for Preparing Key Products[1]

Personnel Costs / Key Products	Total Professional Staff Cost	Professional Cost/ Professional Work	Professional Cost/ Support-type Work	Total Support Staff Cost	Total Key Product Personnel Cost
A	$ 4,026	$ —	$ 4,026	$ —	$ 4,026
B	671	112	559	9	680
C	3,523	—	3,523	—	3,523
D	168	—	168	—	168
E	587	84	503	—	587
F	2,300	449	1,851	18	2,318
G	29,500	14,500	15,000	14,100	43,600
Totals	$40,775	$15,145	$25,630	$14,127	$54,902

Table 3-3 Annual Levels of Effort for Preparing Key Products[1]

Products	Total Professional Staff Effort (Hours)[a]	Professional Effort/ Professional Work (Hours)[a]	Professional Effort Support-type Work (Hours)[a]	Total Support Staff Effort (Hours)[a]	Total Key Product (Hours)[a]
A	288.0	0.0	288.0	0.0	288.0
B	48.0	8.0	40.0	1.0	49.0
C	252.0	0.0	252.0	0.0	252.0
D	12.0	0.0	12.0	0.0	12.0
E	42.0	6.0	36.0	0.0	42.0
F	164.5	32.1	132.4	2.0	166.5
G	860.0	640.0	220.0	463.0	1323.0
Totals	1,666.5	686.1	980.4	466.0	2,132.5

[a] A person-year is defined as 2080 hours in this sample.

52

Physical locations:

   Continent

   Country

   State

   Region

   Department

   Individual work station

Other data classifications may be made by:

   Salary and benefits information by job classification

   Head count projections

   Corporate and local personnel policies and procedures

   Employee turnover ratios

   Temporary and overtime policies and expenses

   Records and file management procedures

Although it is generally difficult to build a data base of this information, such an activity must be started. If it becomes too difficult or expensive to obtain data on a company-wide basis, the focus should be narrowed to a more realistic but representative sample. As the sample size is reduced, however, it becomes more difficult to project and substantiate the findings.

DEVELOPMENT OF AN EQUIPMENT AND SERVICES INVENTORY

Another important segment in the preliminary office systems survey is to take a detailed inventory of existing equipment and services to help classify and identify expense levels and cost-saving opportunities. Chapters 5 and 6 describe how to select and apply the necessary office systems forms to facilitate this effect. To encourage positive user participation, the preliminary datagathering process should be made as simple as possible.

An inventory of equipment and services (see Table 3-4) should include the following categories:

Telecommunications systems and services

Telecommunications networks/facilities

Computer systems and services

Text processing Systems

Calculators

Typewriters

Graphics

Energy systems

Forms design systems

Personal computers

Printing and publication systems

Copier and reprographic facilities

Mail and related distribution systems

Micrographics

Time sharing systems

Library facilities and services

An assessment must also be made of existing procurement and related financial practices, including:

Rental, lease, or purchase procedures and contracts

Inventory category by number, value of equipment, vendor, and location

Procurement: policy and authorization levels

For each category listed in Table 3-4 various statistics should be accumulated: functional utilization and penetration, application classification by new, revised, and maintenance development expenditures and by major expense category (e.g., personnel, hardware, supplies, utilities). Figures 3-1, 3-2, and 3-3 illustrate these statistics for the computer systems component. Similar statistics should be developed for all office equipment and services.

Studies such as those conducted by Booz, Allen & Hamilton and the Stanford Research Institute have clearly revealed that there are significant opportunities to impact personnel, equipment, and ser-

Table 3-4   Office Equipment and Services Expenditures

	($) Current-Year Plan	($) Previous Years	($) Projection	Percent of Change and Reason for Variation
Telecommunications systems and services				
Telecommunications networks facilities				
Computer systems and services				
Text processing systems				
Calculators				
Typewriters				
Graphics				
Energy systems				
Forms design systems				
Personal Computers				
Printing and publications systems				
Copier and reprographic systems				
Mail and related distribution systems				
Micrographics				
Time-sharing systems				
Library facilities and services				
Totals and variance (%)				

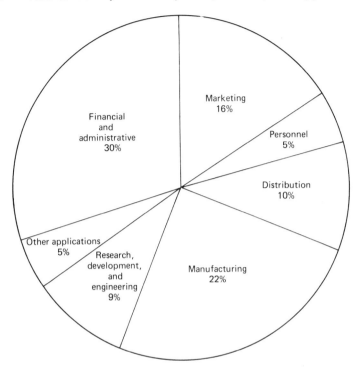

Figure 3-1 Sample allocation of computer systems by function.

vice expense levels through the application of available and future technologies. These studies have projected savings based on detailed analyses of how office workers spend their time, what functions they perform, whether they are employed in structured or unstructured environments, and whether available technologies might affect the related expenses. The studies have also evaluated the direct cost impacts and functional benefits associated with other administrative expenses such as mail, communication facilities, copiers, data processing, travel, office, and use of external office services.

In general, these studies have analyzed specific organizational populations and applications and have extrapolated summary benefit estimates. Estimates for potential average savings have ranged

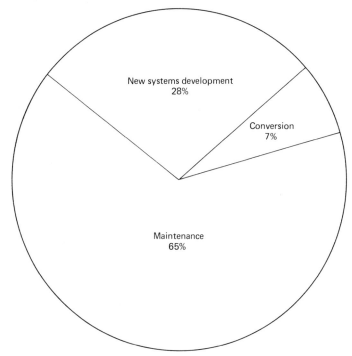

Figure 3-2   Sample allocation of computer systems efforts to new systems, ongoing applications, and conversion.

from 5% to 10% for executive management and from 10% to 20% for managers, professionals, secretaries, and clerks; savings on other administrative expenses (e.g., energy usage) are also in the 10% to 20% range. These industry estimates have been projected on the assumption of widespread office systems implementation. Such generalizations will vary and will be subject to wide variations based on such constraints as industry type, work force composition, type of work performed, level of technical sophistication and business maturity, among other factors.

There are several approaches to evaluating and assessing data on the macro level. One method is to relate cost displacement oppor-

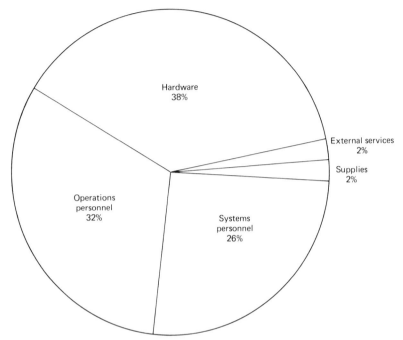

Figure 3-3 Sample allocation of computer systems budget

tunities to entire categories of staff and related expense streams existing within the organization. Yearly savings and costs must be developed for comparative analyses and presentation to management. Some of the costs to be considered include equipment, software, staff, installation and facilities, maintenance, supplies, and communications expense. Training and educational expense requirements should also be calculated. These summary expense estimates may then be multiplied by the number of personnel to arrive at macro expenses, which can then be projected over time, using appropriate financial investment factors (cost of money, return on investment, investment tax credits, etc.). The benefit potentials by personnel can be computed in a similar manner to arrive at a net cost savings range.

With these savings and cost estimates available, capital investment requirements can be projected on a per-capita basis.

IDENTIFYING AREAS OF BENEFIT AND OPPORTUNITY

Some of the benefit and potential savings opportunity areas include:

Personnel productivity/cost displacement opportunity:
  Executive
  Middle and first-line management
  Professional and technical
  Secretarial and clerical
Expanded span of control (greater ratio of management to staff)
Availability of information on a more timely, accurate, and flexi-
  ble basis
Reduction of paper proliferation, filing, and storage systems
Avoidance of increases in staff
Elimination of outside labor and facilities expenses
Reduced travel
Improved use of materials, space, and facilities
Reduced use of external services (postage, messenger, typesetting,
  micrographics, etc.)
Enhanced decision support opportunities
Improved customer service
Improved employee satisfaction levels

The opportunities should be expressed in terms of historical trends
and future projections and should be presented in categories recognizable by senior management (i.e., percentage of sales ratios/ex-
pense, job classification ratios, etc.). Benefits may range from modest to significant, may represent either hard or soft dollar results,
and may be either perceived or real. The key is to present an objec-

tive business investment opportunity that identifies both positive opportunities and negative factors and considerations. The project opportunities uncovered should always be expressed in terms familiar to senior management and should conform to corporate objectives.

Paul Strassman stresses that the sequence for identifying opportunities should start with the budgeting process. It must identify all the components of informationprocessing cost and segment them by (1) *function*, for example, the total cost of performing the billing function, from order entry until receivables are reconciled; (2) *technology*, for example, what portion of the billing cost is done clerically and what portion by computer; and (3) *organization*, for example, what the various billing systems of one organization are.[2]

KEY IDEAS

Preliminary (macro) cost and benefit opportunities may be developed through proper quantification and qualification techniques. Related to this effort, the data developed may be used to secure management approval and funding for an office automation program.

This approach will require a significant and time-consuming effort, and senior management must be committed to the effort.

The assessment activities must include the following steps:

Identification of sources of information

Identification of categories of information to be collected

Development of equipment and service inventories

Identification of benefit and opportunity areas

Extrapolation and development of a range of savings

From the information accumulated, a clear summary of benefits and costs must be developed in a manner familiar to management so as to present an appropriate and acceptable cost displacement and opportunity profile.

Study details and summary results and conclusions should be discussed in clear and concise terms that should be related directly to business objectives, plans, and opportunities. Technical presentations should be avoided at all costs.

REFERENCES

1. *Guidance on Requirements Analysis for Office Automation Systems,* Institute for Computer Sciences and Technology, National Bureau of Standards, Washington, D.C., September 1980.
2. Paul A. Strassman "Managing the Cost of Information," *Harvard Business Review,* September–October 1976.

CHAPTER FOUR

How to Organize the Program?

Before organizational issues can be discussed, it is necessary to first understand office functions, the concept of office automation, how to begin and plan for office automation, and how to identify cost savings opportunities. The organizational effort, however, relies less on systems experience and more on management and administrative abilities. The office systems organization will require a blending of skills capable of developing objectives and performing systems studies, selling the concept, and implementing, coordinating, and monitoring workable and realistic strategies and action programs.

In any endeavor, the right people in the right positions can make the difference between success and failure. If management is serious about office systems and its potential benefits, it will have to commit the necessary manpower and capital. The office systems organization must have multiple skills and abilities, including data processing (hardware and software), text processing, communications, industrial and facilities engineering, administration, human factors, internal marketing, and training. These skills may be purchased from consulting services, borrowed from within, or hired. Without the proper level of experience in each of these areas, the office automation effort will have only limited success.

ORGANIZATION ISSUES

Over time, office systems will become pervasive and will have a significant impact throughout the organization. Before organizational recommendations are presented, some basic questions must be answered in order to establish the appropriate criteria for structuring the office systems department and positioning it within the company. The questions that must be answered include:

1. Is a project, a functional, or a matrix structure best?
2. To which department or division in the company should this function report?
3. Should any existing functions be combined with this new one?
4. How large should the office automation department be?

64

5. Should the department be a totally or partially centralized function, or should it be a decentralized function?
6. What is the corporate role, compared to the operating unit role?

Peter Drucker recommends that before an organization is formalized, the desired objectives and strategies should be delineated. Drucker also suggests that the function should be described as either critical to meeting corporate objectives and therefore essential, or not critical. He further points out that information activities, "unlike most other result-producing activities, are not concerned with one stage of the process, but with the entire process itself."[1] This factor complicates the placement of an office systems department because it implies that the department must be both centralized and decentralized. Therefore, the key question is: How critical is office systems to the organization? The answer to this question will help to determine the placement of office systems in the organization.

ORGANIZATION ALTERNATIVES

Three basic organization structures exist: the project, the functional, and the matrix. The project structure requires that managers supervise particular projects, which can vary in size and duration. The functional structure (e.g., accounting) typically has a manager supervising a specific function and projects within the function. The matrix organization represents a combination of project and functional characteristics and also provides a framework for checks and balances. For instance, corporate staff will often place personnel within other divisions to determine that corporate interests are ensured or that a strong dottedline relationship exists between line and staff functions. According to Jay Galbraith, "The primary design issue in the matrix is where to establish the dual reporting relation in each laboratory, department, etc. This is determined mostly by technological determinants of work divisibility." The inherent problem of duality must be overcome, Galbraith continues, by creating "a power balance between department managers, both of whom champion different sets of goals." Thus, in summary, "the matrix design institutionalizes an adversary system."[2] Selection of the appropriate structure will be contingent on managment attitudes and styles. However, the project structure is typically the best suited for the initiation stage.

OFFICE SYSTEMS CHARTER

Independent of organization structure, the *office systems charter* (OSC) must be scoped and must contain both general and specific objectives and responsibilities. A sample office systems charter follows:

I. *Purpose, Objectives, and Responsibilities*

The purpose of the Office Systems Department is to assist the organization to achieve its short- and long-term objectives through the use and application of more effective office systems techniques and technologies.

Examples of objectives and responsibilities follow:

1. Develop and recommend strategic, tactical, and operational plans.
2. Assess and monitor technology and related issues.
3. Provide advice and counsel to clients.
4. Develop, design, and implement research activities.
5. Hire and develop staff.
6. Establish funding requirements and priorities.
7. Develop standards, policies, and guidelines.
8. Establish an education and communication program.
9. Implement and evaluate pilot systems.
10. Coordinate office technology vendor relations and contracts.
11. Issue office technology systems study and planning guidelines.

II. *Scope*

Office systems consists of, but are not limited to, electronic technologies, including text and data, voice, video, communications, image, and graphics, as well as related processes and services.

ORGANIZATION DEVELOPMENT

The organization of the department requires a range of management skills and different technical specialties as it evolves through various periods of development. The initial or learning period is followed by the planning, implementation, and operations periods. The learning and planning activities will be ongoing, even after the implementation and operating periods have begun. Figure 4-1 illustrates the time frames for each period.

Before the learning period can begin, it will be necessary to assign overall responsibility to a manager, who should have a diverse technical background as well as a strong business orientation. This person should also be sensitive to people needs, be able to communicate well and to market effectively, be analytical, be able to manage projects, be politically astute, and have an entrepreneurial spirit (see Figure 4-2).

The next step is to develop an organization and recruit the required staff. At this point, the office systems department should be part of the corporate staff, regardless of the degree of corporate centralization or decentralization. As learning activities begin, the group should have limited line affiliations and the complete support of management. As the function grows and matures, a corporate presence should be maintained to assure uniformity.

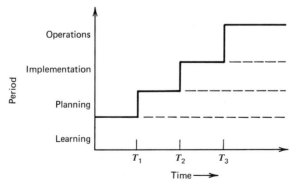

Figure 4-1 Periods of organization development.

POSITION TITLE: Department Head—Director Office Automation
DIVISION/DEPARTMENT: Management Information Systems
MANAGER'S TITLE: Vice President—MIS
NEXT LEVEL OF MANAGEMENT: Division head or group head

PROFESSIONAL REPORTS: 3 NONPROFESSIONAL REPORTS: 1
BUDGET: $X,XXX,XXX LOCATION: City, State

PROJECTS:
Develop a five-year office automation plan.
Evaluate existing and near-term technologies.
Analyze the requirements of professionals and nonprofessionals.
Hire necessary personnel to meet objectives.
Develop guidelines and controls for the acquisition and utilization of
hardware and software systems.
Develop the techniques and guidelines necessary to assess user
requirements.
Educate management and staff in office automation concepts,
programs, and use.

PRINCIPAL RESPONSIBILITIES:
To manage a staff of three professionals and one nonprofessional
with an annual budget of $X,XXX,XXX.
To orient this department and educate staff in the principles and
concepts of office automation. To determine the needs and
requirements throughout the company for office systems in general.
To develop a strategic and operational plan which will include
research and development of pilot projects that will result in
cost-effective systems implementations.
To propose a policy for the acquisition and utilization of hardware
and software systems and to implement controls to ensure policy
adherence.
To assist prospective users in application definition and vendor
evaluation.
To establish vendor relationships and execute nondisclosure
agreements as necessary.

Signature of applicant: _____ Date: _____

Manager's approval: _____ Date: _____

Next level of management approval: _____ Date: _____

Figure 4-2   Position Description—Director Office Automation

68

Figure 4-3 positions the office systems department in management information systems (MIS). Many organizations have already set up similar departments under Communications or Administrative Services.

Management information systems have been evolving with a heavy dependence on data communications, and it is already quite common to find voice and data communications under the same MIS structure. The authors believe that most information services functions will eventually become part of the MIS structure and that MIS will develop into a structure known as information resource management (IRM) (see Figure 4-4). IRM will eventually include such functions as micrographics, administrative services, human factors, and others.

Learning

A number of areas must be studied during the learning period. They include an assessment and understanding of the technology, an awareness of what other organizations and consultants have been doing and are planning, identifying the user community, determining the business requirements, and understanding the associated expenses.

Chapter 9 reviews and describes the technologies. It is critical for the office systems department (OSD) to become familiar with all

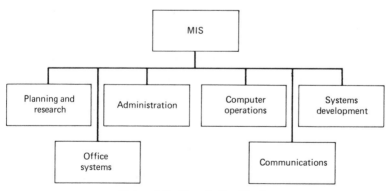

Figure 4-3 Organization Chart.

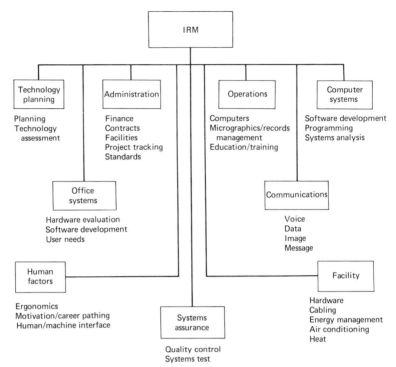

Figure 4-4 IRM organization chart.

these technologies. A number of alternative approaches will accomplish this.

During the learning period, members of the office systems department should be assigned to study the technologies, to visit other organizations, and to determine user and business requirements. Among the skills required for members of the OSD are computer science, communications, systems analysis and design, human factors, facility planning, personnel management, and organizational and operational planning and management. The first employee assigned to this department should have a solid computer systems background, preferably with applications and analysis experience. A sound knowledge of communications would also be very helpful. (See Table A-1 for a sample job description.)

The second member of this team should have communications expertise with some computer systems experience. This person should understand voice and data communications and networking, and should have prior experience in line protocols, modems, store-and-forward message switching, on-line data networks, and telephone (PBX) systems. Familiarity with domestic and international communications carriers, satellite and microwave communications, and services such as Mailgram® would also be helpful (see Table A-2).

The third team member should be a systems analyst. In addition to applications and analysis skills, this person should have strong written and oral communications skills and should have the ability to encourage users to accept change (see Table A-3).

This team must work as a cohesive unit. Such cooperation will result in a sharing of knowledge and a better understanding of common goals and objectives.

Planning

Six to 12 months after formation, the office systems department will be ready to enter a formal planning period. By that time, activities undertaken during the learning period will have resulted in a solid reference base from which to proceed.

As the team develops long- and short-range plans, it must consider how the existing and developing technologies can be best used throughout the organization. The potential effect of these technologies on human resources, facilities, and procedures must also be determined—a complex process that may require special internal and external consulting expertise and resources. In many companies, senior management often depends upon the recommendations of well-established outside consulting firms.

At this point, consideration should be given to adding three more staff members or to acquiring the necessary skills from other departments on an as-needed basis. The first additional member should be a human resources specialist whose tasks are to evaluate current organizational structures and procedures and to determine what changes may result or may be required through the widespread adoption of office systems. This person should be familiar with organizational theories and methods: turnover ratios and related causes, work relationships, societal needs, ratio of staff to line employees, ratio of managers to subordinates, the various demographic trends in white collar professional and nonprofessional growth and associated expenses, the increasing difficulty in finding and keeping qualified support staff, and the expansion of white collar unionization (see Table A-4).

After a thorough analysis of the technology, it will become clear that the implementation of electronic systems will have a significant effect on the physical environment. These systems require electric power and cabling, and they generate heat and noise; therefore their physical appearance (e.g., size, color) must be considered. It is crucial to plan adequately for future installations in existing as well as in new office buildings. Careful planning in other areas can be totally negated by inadequate facilities planning.

The next addition to the staff should be a facilities specialist who has had experience with space and electrical planning, ergonomics, air-conditioning and ventilation equipment, relocation, security, and office design. This person should consider and evaluate the heat, vibration, and noise potential of new technology and should make appropriate recommendations. Additional areas include static electricity conditions, furniture, installation coordination, and relocation (see Table A-5).

The third addition, a person with administrative services experience, may be appropriate at this time. This person must be familiar with the development and distribution of policies and procedures, must have a solid understanding of the various information gathering and processing procedures, and must be able to help evaluate the potential of new electronic tools on all these processes.

At this time, other departments in the organization may have to be consulted for additional information. The office systems department must meet with the corporate auditing department to determine the levels of control and security required for office systems. It is critical to discuss strategies with staff and line management to ensure that the objectives of all units are similar and compatible. It is important to communicate objectives and status to the management of other business units and to seek their acceptance, cooperation, and support. This communication will prove valuable during the implementation period. Figure 4-5 shows the OSD department during the planning stage.

Implementation

Once the office systems plan is approved, it must be implemented. In practice, however, the implementation period should begin before the plan is actually approved (this is a minimal risk approach and provides the planner with more time for implementation).

The implementation program will require decisions on schedules, prospective users, vendors, resources, priorities, scope, applications, locations, tests, and procedures. Because an implementation effort will affect the company's ongoing business and existing practices and procedures, extreme care should be taken during this period. It is during the first few systems installations that the success or failure of the office systems department will be measured. The first installations will be observed not only to prove the validity of these concepts, but to prove the effectiveness of the office

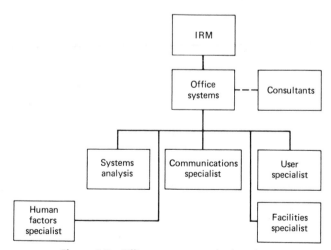

Figure 4-5 Office systems organization chart.

systems department's management and staff. The OSD head should be as concerned with the implementation and operations period as with the learning and planning periods. (More on implementation is covered in Chapter 7.)

An additional full-time staff member will be needed to support the training requirements. This person will train users, write manuals, provide general education, and instill self-confidence in users.

As the organization accepts office systems and begins widespread implementation, it may be desirable to develop an in-house training staff capable of not only initial training and post-installation support but all training requirements. The objectives will be to train on the equipment as well as to develop appropriate organizational procedures. Thus the organization can set and achieve higher standards. Furthermore, the authors are aware, regretfully, that many vendors do their best (and perhaps their *only*) training when a system is initially installed. Once the system is in place, vendor support often tends to diminish; for this reason internal training is essential.

It is highly unlikely that the size of the office systems group will be allowed to increase further at this time. Once several successes have been demonstrated, however, this situation will change. The skills of the office systems staff will be diverted from full-time planning to a combination of planning and implementation. The systems and communications specialists will become heavily involved in developing functional specifications for recommended systems. The specifications should include features that are required to accommodate user needs. The user specialist should review all specifications to ensure that requirements are met. The facilities planner must be certain that all selected locations can be serviced and that they are provided with proper power, communications support, space, and ventilation.

Once a systems specification is approved, the department will occasionally submit requests for proposal (RFP) to vendors. A vendor's products must be evaluated, as well as its reputation, financial soundness, and overall customer service capability. The office systems department must fully test and evaluate new systems before installation in the user environment, and particular emphasis must be placed on the testing of new products.

Operations

Before equipment is installed, systems must be developed that are capable of directing and monitoring equipment and software performance. It is essential to provide the best possible user support for all office systems. A centralized trouble reporting system in which user questions can be answered, calls can be recorded, vendors' service personnel can check in, and operations personnel can track systems malfunctions should be established. Initially, when the equipment base is small, the training specialist can easily answer most operational questions; as more and more electronic systems are implemented, however, a separate operations function may be necessary. This revised structure is illustrated in Figure 4-6.

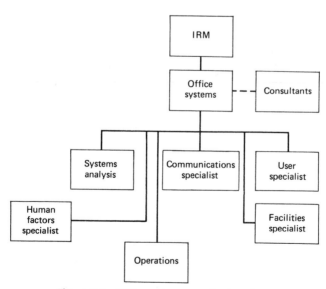

Figure 4-6   Office systems organization chart.

ORGANIZATION PERIODS WITHIN THE STAGES OF DEVELOPMENT

The four periods of organization development outlined above fall within the stages of office systems evolution outlined in Chapter 1 (see Figure 4-7). From an organizational perspective, the learning and planning periods occur during the same period as the initiation stage. The implementation period occurs during the expansion stage; however, the expansion stage extends beyond implementation into the operations period. Systems must be installed and must be operational before the organization can enter the formalization stage. Integration and maturity depend on a critical mass of systems being in place and on technological compatibility among these systems.

Throughout these development periods, corporate guidance and direction are vital. It will prove too costly and unwieldy for each operating entity to establish its own direction without regard for company-wide office systems integration. Each operating division can determine its own need, but the solution, if it involves automation, must be implemented within established corporate objectives, guidelines, and strategies. Such a situation could exist within a multidivision corporation whose objective is to improve business communications with an electronic mail network. Each division in the corporation has its own objectives. Without central coordination, the company might end up with multiple, incompatible electronic mail networks instead of one corporate-wide, mail service.

RECOMMENDED OFFICE AUTOMATION ORGANIZATION

A centralized organization requires an organization similar to the one shown in Figure 4-5. If a function of the office systems department is to provide a consulting service for internal users, the systems and user specialist units will expand accordingly. In order to ensure the productive utilization of these efforts, it is suggested that the office systems department charge its internal clients for time-

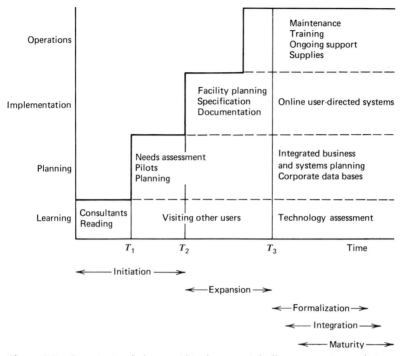

Figure 4-7 Organizational phases within the stages of office automation evolution.

and project-related expenses. If the office systems department will also be providing implementation, training, and operations support to internal users, the implementation and operation units must expand and must charge clients as appropriate.

Given a decentralized organization, it is more likely that the department outline in Figure 4-8 will be requried. This will be especially true if the operating groups develop their own departments and interface directly with corporate office systems. Whatever can be accomplished at the division level—for example, needs studies or training—should be supported at that level. The functions of corporate planning, technology assessment, overall coordination, and prototype evaluation should be performed at the corporate level.

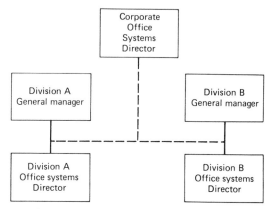

Figure 4-8   Decentralized coordination chart.

According to Derek Newman, "Expansion of work or greater delegation of autonomy towards the periphery must be balanced by adequate control from the centre if decentralization is not to lead to incoherence by default."[3] Conflicts over control are certain to arise since decentralization requires the decentralization of decision making, authority, and risk taking, along with the process of evaluation. This situation is compounded because the expectations and management styles are often different among divisions, subsidiaries, and headquarters.

KEY IDEAS

In this chapter the reader has been familiarized with various organization alternatives. There is no unique approach. Each company must first understand its own objectives, commitments, politics, and style before a "solution" can even be attempted. There are many and varied organizational strategies and alternatives, and the factors described in this chapter, combined with past experiences, will provide the reader with a framework from which to proceed.

Wherever the office systems department is positioned and regardless of whether it has a centralized or decentralized organization structure, the concepts outlined in this chapter are valid. Expertise in computers, communications, applications, systems, facility planning, training, human factors, operations, and personnel is a requirement for success. Most organizations will start small and evolve proportionately, depending on their level of success. Eventually, the office systems department will take on a traditional data-processing structure, including research, planning, development, systems assurance, administration, implementation, and operations. In practice, the term "office automation" should be replaced by office systems and should become less and less isolated, ominous, or mysterious.

Office systems may be one of the newest and most dynamic growth areas of an overall IRM philosophy. Proper organization and placement, and careful staffing, will both facilitate this growth and ensure ultimate success.

REFERENCES

1. Peter F. Drucker, *Management* (New York: Harper & Row, 1974), pp. 529–549.
2. Jay R. Galbraith, *Organization Design* (Reading, Mass.: Addison-Wesley, 1977), p. 163.
3. Derek Newman, *Organization Design* (Philadelphia: International Ideas, 1973), p. 91.

CHAPTER FIVE

How to Coordinate and Control?

One widely accepted definition of management control has been provided by Robert Anthony: "Management control is the process by which managers assure that resources are obtained and used effectively and efficiently in the accomplishment of the organization's objectives."[1]

There are many degrees of control, ranging from tight centralization to loose decentralized coordination. Control depends on factors such as formal organizational policy and responsibility, the degree and style of management centralization or decentralization, the corporate role (strong or weak), the degree of geographic dispersion, approval of funding or commitment levels, and the status of the particular stage of office systems evolution as described in Chapter 1.

Control encompasses coordination, monitoring, and linkages with the planning and budgetary processes. In most large organizations, it is impractical to evolve control systems that have the total support of the entire organization. However, by applying the appropriate plans, policies, standards, guidelines, and coordination techniques, a broad consensus of the overall objectives and direction of the office systems efforts can be attained.

COORDINATION AND CONTROL IN CENTRALIZED, DECENTRALIZED, AND MATRIX ORGANIZATIONS

Before a particular coordination and control strategy can be formulated, the organization as a whole must be studied and fully understood. For example, the holding company structure, which is basically an umbrella structure, represents the extreme form of decentralization. In an organization that is divided into several major and autonomous business units, the corporate role is extremely weak and virtually nonexistent, except perhaps for tax-reporting and consolidation purposes. In this example, corporate office systems coordination would be practically nonexistent and every business unit would be independent.

Another variation of structure would be a centralized organization with major divisions and subsidiaries and a significant centralized corporate support function. Under such an organization, a central office systems department would exist with a significant amount of leverage. Its function would be to consult with and to direct, control, or influence the various divisions and subsidiaries. This may require a matrix form of organization in which overall corporate direction and policy are established by a corporate-based office systems department, but which permits the business units a high degree of freedom as long as they operate within the corporate guidelines.

Richard Hall points out that

> "the major defining characteristic of centralization is that power distributions are determined in advance by the organization. Centralization thus can refer to either individuals or units, such as divisions or departments, within the organization. It also, obviously, refers to levels within organizations, as when it is specified that only people of a particular rank have the right to make certain kinds of decisions.[2]

A combination of centralized and decentralized organizational characteristics is known as matrix management. As Kenneth Knight has asserted:

> Conflict becomes dysfunctional when it delays decision-making, reduces significantly the amount of energy available for relevant work, blocks important communication channels and subjects individuals to unacceptable stress. The success to which these things will happen could depend on how well a system of matrix management can be integrated with the existing organizational culture.[3]

Participative management and management support are the keys to success in all these various structures. Other management difficulties aside, corporate guidance and direction are imperative. While each business unit must determine its own needs and requirements, these needs must fit in with overall corporate goals, especially where automation is involved.

RESPONSIBILITY, CONTROL, AND THE CORPORATE ROLE

Because office systems have been fractionalized, the control process must be so designed as to overcome issues of territoriality, duplication of staff, and operational compatibility. A key factor in the successful control and coordination of the office systems activities is the precise role defined for the corporate office systems department.

The role of the corporate department responsible for office systems will vary according to the environment. Table 5-1 illustrates several important characteristics of office systems in centralized, matrix, and decentralized organizations, with particular emphasis on the role of the corporate office systems department. Although the table describes the extreme forms of the corporate role, the role will vary depending on the management incumbents and their styles. The selected organizational alternative will have a significant impact on the degree of control that the corporate office systems department has over such activities as short- and long-range planning, project management, systems development, vendor selection and equipment, economies of scale, company-wide uniformity, compatibility, and training at various levels. However, independent of a particular environment, a company-wide coordinating role for the corporate office systems department will help to overcome the constraints suggested by the phases and stages of office automation evolution described in Chapter 1.

Even though individual management philosophies will dictate the alternative selected, most organizations will choose the matrix alternative as the most practical and realistic for the following reasons:

1. Provides coordinated guidance and direction while encouraging decentralized operating responsibilities.

2. Reduces duplication of effort and resources over a fully decentralized environment.

3. Assures that the appropriate administrative methods, cost controls, authorizations, standards, and guidelines are in place.

4. Provides corporate expertise, guidance, and advice in applying state-of-the-art concepts.

5. Shares research, prototype, and actual experiences as well as the costs of developments, while spreading the risks at the same time.

6. Stimulates communications, back-up, and other linkages between semi-autonomous business units.

Thus, the matrix alternative preserves the autonomy of each major operating entity and helps each organization in developing office systems that optimize the use and allocation of resources. Limitations of various organizational alternatives are:

Decentralized alternative. Costs in the decentralized environment are higher. Decentralization without direction or control results in extensive duplication of systems development, personnel, equipment, facilities, costs, and other efforts. Integration is extremely difficult if not impossible to achieve in this environment because of incompatible hardware, software, communications protocols, and the general political environment.

Matrix alternative. Because of built-in checks and balances, this approach may cause conflict and polarization, may require excessive time to make decisions, and in terms of overhead expense, is generally less costly than decentralization but more costly than centralization.

Centralization alternative. With many corporate management philosophies trending toward profit decentralization and participative management, and with the rapidly accelerating concept of distributed, shared, and microminiaturized technology, this alternative is becoming increasingly difficult to develop and maintain.

OFFICE SYSTEMS PROGRAM REQUISITES

Plans, and the resultant budgets and objectives, provide a framework for monitoring progress and represent the mechanisms for measurement, coordination, and, ultimately, control.

Table 5-1 Office Systems Organizational Alternatives, the Corporate Role, and Other Characteristics

Organizational Alternatives	Office Systems Department	Office Systems Equipment/ Facilities	Application Development	Resource Sharing	Corporate Role
Decentralized	Highly independent and autonomous business units	Multiple vendors Compatible by chance only	Local development	None	Virtually no influence on business units and other areas Micro-perspective
Matrix	Semi-autonomous—report to business unit management	Independent, with limited sharing and compatibility Multiple	Local development, with common development when	Some sharing	Active role in planning, equipment acquisition, common

Centralized	Dotted line to corporate office	vendors	practical		applications, research, and education Consultative role for corporate staff business units Checks and balances
	Central control and authority	Limited vendor equipment Strong standards and guidelines Highly compatible	Developed at corporate level	Maximum sharing Economies of scale	Dominant role in planning, implementation, and control

To assure the implementation of the proper office systems coordination and monitoring activities, the formal program should include the following requisites:

Charter. Describes mission, responsibilities, authority, and relationship of the corporate office systems department with the business units (and their divisions and subsidiaries) and with other corporate departments. A sample charter was described in Chapter 4.

Policy statement. Provides authority for office systems directions.

Procedures/standards/guidelines. Directs office systems programs covering such areas as: planning, application descriptions and development, hardware/software selection, compatibility, transmission protocols, study methodologies, management issues, cost and benefits analysis, security, and other areas (see Table 5-2).

Project management systems. Procedures should be in place for documenting projects, estimating costs, allocating resources, project planning and scheduling, benefits identification, proper authorizations, and contingency plans.

Office systems management planning steering committee. Achieves senior management involvement and commitment to the program and insures the relevance and linkage of office systems plans to business plans, objectives, and directions.

Planning

1. *Annual office systems plan.* Supports expense and capital budget, personnel, and other resources; Should include primary and secondary goals and relate long-range and short-range plans.

2. *Long-range office systems plan.* Supports strategic business and systems plans over a longer time period.

Periodic progress reporting. Establishes project reporting mechanism for status, resource utilization, trouble, and benefit realization reporting.

Table 5-2   Suggested Office Systems Policies, Procedures, Guidelines, and Standards

Type	Subject
Policy	Planning authority Responsibility/charter Privacy Funding and expenditure authority
Procedure	Request for use of office systems resources Project management and reporting Emergency and contingency plans and back-up strategies Documentation Inventory of hardware, software, services and related communications resources Project audit and realization Installation checklist Utilization of outside services security
Guidelines	Office systems development methodology (OSDM) Vendor selection criteria/questionaires (see Appendix D) Request for proposal (see Appendix D) Microfilm systems—evaluation and administration Word processing systems—evaluation and administration Reprographic systems—evaluation and administration Communications systems—evaluation and administration Multifunctional work stations (terminal)—evaluation and selection Survey of needs and requirements Education courses and newsletter Planning
Standards	Glossary of terms Standards for data communications and other media integration/interface Data base/data dictionary

These are some of the formal requisites critical for the establishment of office systems program controls. In addition to the formal program requisites listed above, a number of coordination strategies should be considered. They include the following:

Management council. This council consists of nontechnical management and other key personnel from various corporate staff functions and business units to provide ideas and insights with respect to the future needs and applicability of office systems technologies within their environments.

Technical council. This council consists of technical representatives from each business unit's IRM department to provide a forum for sharing and discussing current and future technical and operating issues, trends, and techniques.

Periodic review visits to business units. The corporate office systems department should meet periodically with the business units to review progress, problems, and other matters of common interest.

Company-wide office systems meeting. Periodically, all of the managing staff of the company-wide office systems program should meet to discuss areas of common interest, conduct technology briefings and trends, and be updated on office systems strategies and programs. Speakers from both the company and outside should be invited to provide a broad basis of stimulation.

Regular progress reports. Progress reports reflecting corporate and business unit projects should be submitted by all groups and circulated by the corporate office systems department.

Training and education. The corporate office systems department should sponsor and/or conduct various training and education programs for technical and nontechnical personnel on a company-wide basis.

Internal consultants. Internal corporate office systems consultants should provide advice, counsel, and guidance within the company on a periodic or on an as-requested basis.

Internal research studies and reports. The corporate office systems department should conduct and distribute on a regular basis research reports on related areas.

Prototype funding. The corporate office systems department should provide funding for and coordination of prototype projects. Pilots may be implemented at the corporate unit or at the business unit. Joint funding of protoypes is also a viable strategy to encourage greater involvement and commitment between corporate and business units.

Newsletter. A newsletter should be published on a regular basis. Contributions from all interested parties throughout the company should be requested and encouraged.

Career pathing. A program should be established between the office systems and the corporate and local personnel departments to establish coordinated career paths for office systems personnel.

User forums and suggestion program. A periodic forum for user personnel to share their ideas, experiences, and problems should be encouraged.

Technology caravans. Vendor presentations and equipment demonstrations should be scheduled internally on a periodic basis.

LINKAGE OF PLANNING, DEVELOPMENT, AND CONTROL PROCESSES

Many organizations do not consciously interface the various elements of the planning, development, and control processes. To effectively coordinate and control an office systems program, strong linkages must be established between the processes.

Figure 5-1 illustrates the structure and relationships of the office systems planning, development, and control processes and components in terms of (I) goal setting, problem identification, and intelligence gathering activities (e.g., external and internal environmental scans, business and systems plan objectives), and assumptions, (II)

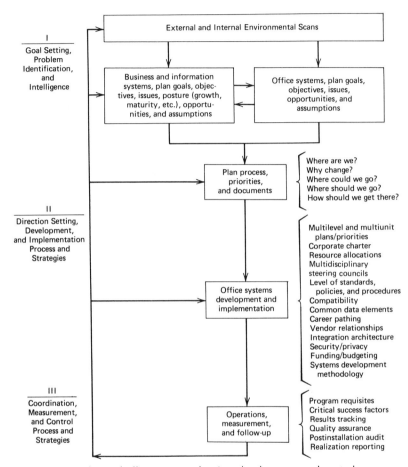

Figure 5-1   Linkage of office systems planning, development, and control processes.

development and direction-setting strategies (e.g., systems development methodology, standards and guidelines, vendor strategies, resource allocations), and (III) coordination and control activities (e.g., measurement criteria, program requisites, quality assurance).

It is vital that the coordination and control mechanisms be linked with and provide feedback into the planning and development processes on a continuous and interactive basis to maximize effectiveness and to minimize problems and risks.

KEY IDEAS

Control and coordination of the office systems activities is a difficult process because of their broad and complex dimensions. This chapter has discussed several formal and informal control and coordination mechanisms. A blending of both mechanisms under a modified matrix management concept will probably be required in most organizations.

Certain principles should be followed in coordinating an office systems program. Some major principles include the following:

Develop senior-level management steering committees and advisory councils at various levels to obtain approval, involvement, and commitment.

Establish office systems program requisites.

Link the office systems plans and activities with the business and information systems plans, development cycles, budgets, and control processes.

Establish a corporate office systems department with a specific charter and mission.

Create a climate congenial to office systems coordination activities and not resistant to them.

Make sure that top management and business unit managers really understand the nature, scope, risks, and coordination strategies involved with and required for office systems penetration.

Establish controls to ensure that office systems does not run wild.

In summary, the emerging and interrelated areas of data processing, telecommunications, and office technology have already created some overlap among traditional organizational functions. The need for integrated planning and coordination will require effective leadership, change agents, and motivators. Close coordination among data processing, office systems, administrative services, and the user is a prerequisite for success in any type of organization.

REFERENCES

1. Robert N. Anthony, *Planning and Control Systems—A Framework for Analysis* (Boston: Division of Research, Graduate School of Business Administration, Harvard University, 1965), p. 17.

2. Richard Hall, *Organizations: Structure and Process* (Englewood Cliffs, N.J.: Prentice-Hall, 1975), p. 181.

3. Kenneth Knight, *Matrix Management* (New York: Gower Press, 1977), p. 165.

Part Three

ANALYSIS, REQUIREMENTS AND IMPLEMENTATION

CHAPTER SIX

How to Analyze Needs and Requirements?

Any savings must be measurable and tangible over time, whether they include cost avoidance, cost reduction, improved return on investment, expanded capabilities, or improvement in the quality of work. However, few analytical frameworks have been developed for an understanding of the processes and costs within an office or administrative function. The flow and volume of information, intra- and interfunctional communications patterns, the time required to complete various administrative tasks, and the way in which people work and utilize equipment have not been carefully studied, analyzed, or documented.

Many of the scientific management principles that apply to the industrial sector or factory, such as time and motion studies, work breakdown (specialization) structures, and process flows, also apply in the office environment.[1]

As in the factory, the key to office systems analysis is the identification and analysis of the processes and procedures followed. To establish a sound analytical base, it is necessary to understand that most office workers spend much of their time following relatively unstructured processes. Most office activities can be classified within a broad framework, which can be analyzed in a logical sequence through the use of tools such as transactional analysis, input and output models, flowcharts, decision tables, and work and time flow measurement techniques. Such a framework must include the following general steps:

Break down a job. List and observe all the components (volumes, processes, times, contacts, interruptions, activities, etc.) as they are performed. Be sure to include a representative sample over time so as not to distort the analysis.

Question every detail of the existing job. Why is it necessary? What is its purpose? How can it be performed better? Where else could it be done? Is it being done on a timely basis? What is its value (value analysis)? By whom and for whom is it done? How does it flow?

Determine whether the current processes can be improved, or assess alternative methods of performing the job. Eliminate redundancies, consoli-

98

date functions, simplify work flow and procedures, examine alternatives.

Make recommendations. Obtain the necessary management approvals to proceed with the recommendations and subsequent action program.

Although the preceding framework revolves around a job or a process, the identical framework can be used at the department, division, or corporate level. As already suggested, while the analysis of office and administrative processes may not necessarily lead to office systems, it may lead to significant improvements in the manual processes. Whether an office process is perceived as being either structured or unstructured, it must still be analyzed. Although some processes may not lend themselves to automation, the analysis itself may uncover savings opportunities that were not previously anticipated or recognized in the existing processes.

The authors believe, however, that without this type of analysis, little if any effective automation will occur on any level in an organization. Therefore, if the analysis is properly conducted and sponsored, the results will provide management with decision-assisting information for initiating projects and selecting high-payoff office systems opportunities.

OVERVIEW OF THE TOOLS AND METHODOLOGIES

A study methodology must be developed to utilize the analytical framework effectively. Whatever tools and techniques are employed in the conduct of an in-house office and administrative study, an overall methodology that identifies the logical sequence of analytical activities must be developed. This will assure the establishment of a uniform and standard methodology that can be used by multiple groups in different locations or divisions throughout an organization. The study methodology should consist of the following sequential tasks:

Establish objectives (conclusions that require validation). Understand the administrative costs and bottlenecks; establish a reference base for cost-saving opportunities; focus on a needs and requirements assessment that covers corporate and business unit functions at multiple job levels.

Identify the scope of the effort. Identify areas to be studied or not to be studied (e.g., company wide, number of business units, number of departments, number and levels of personnel, and others).

Develop a detailed study work plan. Identify the study phases, major milestones and overall schedule, resource requirements, responsibilities, and deliverables.

Identify the study team. Select study team members on the basis of survey requirements and scope. Determine the need for external consulting support. Assure that a charter or other communiqué approving the effort has been issued.

Determine data and information to be collected. Spend time on relevant activities: mail flow, copier statistics, telephone usage, filing volumes and patterns, micrographic usage, equipment inventories, data/image communications usage, and average salaries by job classifications. Determine security and confidentiality of data, superior/subordinate ratios, principal/secretarial and clerical support ratios, indirect/direct expense ratios, and others.

Determine how to obtain data and information. In order of reliability, some popular techniques for data collection are: structured observation, diaries, questionnaires, and interviews. Structured observation is the most reliable technique, but it is also the most costly and time-consuming. Interviews tend to be the most unreliable and subjective technique. Therefore, a combination of the four methods should be considered, depending on the specific environment to be analyzed.

Develop and design data collection, analysis, and validation tools and procedures. This will vary depending on individual tasks, job types, organization hierarchy, physical environment, industry, office population, and culture. An assortment of tools and procedures should be prepared:

    1.   Questionnaires (e.g., general company data, principals, secretaries, special-purpose).

2. Observation checklists (of information and data to be collected).

3. Interview guides, diaries, and forms (e.g., to record such activities as mail flow, telephone calls made or received, typing, filing, distribution of time by key tasks and activities)

4. Determine methods for analyzing and processing the survey results (e.g., several service bureaus can assist in this area or an internal computer program may be designed).

5. Select proper sample size (e.g., stratify by job function, level, and business unit to ensure adequate coverage).

6. Determine validation techniques to verify the study's findings (e.g., use management science personnel to assist in establishing sampling and validation techniques).

Communicate the study's nature beforehand. Correspondence should be distributed and meetings should be held with the managers and employees to be involved in the study, to explain the purpose of the survey and the tools and techniques to be used.

Appoint department/division coordinators. Appoint user department coordinators to work with the study team and to represent their respective departments. This will help to establish credibility, reduce the natural resistance to such study activities, facilitate departmental management sponsorship and approval, and enhance the implementation process.

Conduct and administer the study. Establish the appropriate study controls, and document findings (e.g., volumes, times, costs, opportunities, obstacles, flows, organization).

Collect, process, consolidate, and analyze data. The results (e.g., work load distribution, ratio of principals to secretaries, filing storage represented as square footage per employee, typing throughput, internal and external communications, type of correspondence) should be collected, consolidated, and analyzed. Focus on major opportunities, bottlenecks, and limitations.

Present preliminary findings to business unit/functional department managers on an individual basis. To confirm or modify initial findings, enhance communications and obtain continual commitment on a one-on-one basis. This task is critical for gaining credibility prior to a senior management presentation.

Prepare and issue a report and presentation to senior management. Present findings and recommendations to management for concurrence, assigning of priorities, capital and resource commitment, and approval to proceed.

Implement the action program. Carry out the recommendations to completion.

The preceding material has provided an overview of the steps and components required to conduct an administrative study. The remainder of this chapter focuses on defining and exemplifying the data collection and analysis methodologies and discusses some pitfalls to be avoided.

KEY DATA ELEMENTS, SAMPLE TOOLS, AND DATA-GATHERING FORMS

The design and administration of the data collection phase is one of the most critical activities in the preparation of an office study. Generally, the design of the data collection tools and guidelines will be a function of one or more of the following factors:

1. Number and job mix of the population to be included.
2. Job levels to be covered.
3. Number of business units, departments, and locations; their geographic distribution.
4. Costs of the study.
5. Degree of management support and sponsorship.
6. Availability of computers to assist in processing, consolidating, and reporting the results.
7. Sample sizes and validation techniques.
8. Overall objectives for conducting the study in the first place.

Although specific data collection methodologies may vary from company to company, a generic set of analysis techniques should apply to most office and administrative functions. Examples of the techniques and related checklists, questionnaires, and forms for a broad range of office activities are provided in this chapter and in the Appendixes. The emphasis of this section is on the overall data collection effort and the establishment of the appropriate analytical framework. In general, excessive detail has been avoided, especially with respect to the manner in which various techniques and guidelines are to be executed.

The data collection framework for the office environment must focus on a number of areas, which include the following categories:

1. Level and size of audience to be analyzed:
   (a) Principals:
       Senior management
       Middle management
       Lower management
       Professionals
       Technicians
   (b) Support staff:
       Secretaries
       Clerks
       Paraprofessionals
2. Areas to be analyzed:
   (a) General management:
       Corporate level
       Divisional level
       Plant level
       Other levels
   (b) Functional areas:
       Finance (includes Treasurer and Controller)
       Manufacturing
       Marketing

Human Resources

Engineering and Research

Government Affairs

Public Affairs

Distribution

Management Information Systems

Strategic Planning

Environmental Planning

Purchasing

Other

3. Processes or activities to be analyzed which are hierarchical, such as:

Task—a basic process.

Job—a collection of tasks.

Group—a network of interpersonal relations based on work flow, authority structure, and responsibilities.

Organization—a combination of one or more vertically or horizontally oriented groups.

Table 6-1 provides a list of 20 office processes and their associated tasks. On the basis of the major factors previously described, it is easy to visualize a number of different data collection tools that will be required to accommodate all the environments and functions to be analyzed. By itself, this topic can be the subject of a separate book. Therefore, the focus of the remainder of this chapter is on a generic set of data collection forms, tools, and methodologies.

Design and Administration

Before a study is conducted, the study team must develop, or must be provided with, an itemized checklist of all the tasks that will be performed as part of the study effort. This list will include such tasks as developing an overall timetable, identifying initial and subsequent information requirements, and developing the questionnaires, logs, and interview guides and related tasks. Table 6-2

Table 6-1 Data Collection Checklist of Office Processes and Related Tasks

Processes	Tasks
Calculating	✓Perform calculations, create graphs and charts, analyze statistical data, recap and summarize
Copying	Copying, on-site
	✓Copying, off-site
	✓Copy center locations
Distributing	Collate by hand
	Bind or staple
	Address envelope or route slip
	Mark name of recipient on copy
	Stuff
	Deliver
	Telecopy
Errands and reception	✓To other departments
	Out of building
	Coffee
	Personal
	Greet visitors
	Escort
	Arrange for property passes
Personnel related	Process time sheets
	Record attendance
	Vacation schedules
	Transfers, relocations, and promotions
	Appraisals
	Career counseling forms
	Skills update sheet
	Consultant record keeping
	Salary files
Record keeping	Post information
	✓Fill out forms
	Track forms
	Use standard reference books
	Create recaps and summaries
	Update organization charts
	Update manuals
	Produce formats
	Education enrollment forms

Table 6-1 *(Continued)*

Processes	Tasks
	Update and maintain budget and financial records
Research and analysis	Read in office (internal/external)
	Look up information
	Use files and periodicals
	Synthesize and recap
	Create reports
	Program archives
	Standards manuals
	Microfilm
Review and follow-up	Maintain follow-up file
	Review reports for special information
	Follow-up with others
Scheduling meetings	Establish date and time
	Adjust
	Reserve room
	Refreshments
	Contact attendees
	Audio/visual equipment
	Schedule superiors/subordinates/peers/external
Appointments	Check calendar
	Update
	Confirm
	Reschedule
Travel	Reservations
	Create agenda
	Maintain records
	Arrange for advance payment
	Expense statements
	Follow-up
	Obtain cash
	Trip reports
Mail, incoming	Leave area, pick up centrally
	Leave desk, pick up locally
	Deliver to in-box
	Open
	Date stamp
	Log

106

Table 6-1 *(Continued)*

Processes	Tasks
	Prepare route slip
	Deliver to local drop
	Deliver to central station
	Leave in out-box
Mail, outgoing	Address
	Fold
	Stuff
	Stamp
	Deliver to addressee
	Deliver to mailroom
	Deliver to post office
Transmit	Mail
	Telex
	Facsimile
	Message switching center
Supplies	Requisition
	Order
	Locate and pick up special items
	Supply room
Typing and	Type from longhand
transcription	Type from shorthand
(typewriter, word	Type from machine dictation
processor, etc.)	Self typing
	Revisions
	Assemble standard text
	Forms
Proofreading	Spelling
	Grammar
	Format
	Statistical verification
Steno/dictation	Shorthand
	Machine
Filing	Prepare index/filing scheme
	Prepare folders
	File
	Retrieve
	Purge
	Develop system
	Bind printouts

107

Table 6-1 *(Continued)*

Processes	Tasks
Creation	Label printouts and binders Organize printouts Sign-off documents Life cycle documents Records retention Handwrite Dictate Use a terminal Talk/telephone Think

provides a checklist of items that must be prepared and/or considered. In starting a study, the team should assume a limited knowledge about the organization and the environment that is being analyzed. The team will thus be forced to use an adaptive or step-by-step approach by which both vertical data (e.g., intrafunctional from the lowest position to the highest position) and horizontal or cross-sectional data (interfunctional) are gathered.

Questionnaires, Interviews, and Logs

It is good practice to design a questionnaire in a time-phased sequence. For example, the first section should deal with information about a principal's or a secretary's current job (as it is known) in terms of activities, contacts, time requirements, and information about the communications (e.g., information received and documents created). The second section should deal with open-ended ideas about how a better system might increase the effectiveness of the job holder, the department, and/or the organization. Areas to be covered should include job responsibilities, organizational structure, delegable tasks and their time and/or dollar values, information needs and requirements, conceptual systems design, and potential benefits. Although questionnaires should be tailored to specific jobs, general samples of

Table 6-2   Data Gathering Checklist

Develop timetable and schedule
Preplanning information requirements
    Company orientation
    Develop objectives and select audience
    Information needed to tailor tools
    Information needed before data entry questionnaires
    Information needed before conducting interviews
General company information
    Business characteristics and profile
    Classifications of principals and nonprincipals
    Salary ranges—principals and nonprincipals
    Trend data (historic and future)
    Total number of employees
    Administrative expenses (indirect)
    Equipment inventory (all office equipment)
    Equipment costs
    Purchase/lease/rent status of equipment
    Cost of office space per square foot
    Employees working at home
    Messenger services
    Filing volumes
    Mail services
    Communications
    TWX® usage
    Manuals—policy and procedures
Questionnaires
    Printing the forms
    Guide for administering the questionnaires
    Editing the principal questionnaire
    Editing the nonprincipal questionnaire
    Questionnaire forms
    Coding the questionnaires
    Data entry of the questionnaires
    Processing the questionnaires
    Questionnaire analysis
Interviews
    Guide for administering the interviews
    Interview selection—sample size
    Principal interview forms
    Executive
    Lower/middle/senior managers
    Professionals and technicians

Table 6-2 *(Continued)*

Secretarial interview forms
Clerical interview forms
Other information gathering tools
Logs—self-recording
Observation/logs
Filed document samplings
Manuals—interviews
Log forms and instructions
    Typing log instruction
    Typing log
    Typewriter line counter
    Action paper sample
    Sample typed documents
    Typing log follow-up interview

    Incoming/outgoing mail log instruction
    Incoming/outgoing mail log

    Telecopier log
    Telecopier interview

    Copier log
    Copier interview

    File sampling inventory instructions
    File cabinet survey form
    File document sampling form

    Manuals—principals
    Manuals—nonprincipals
    Organization charts
    Facility layouts (blueprints)

---

typical principal, secretarial, and other questionnaires are provided in Appendix C.

Approximately 10% to 25% of the questionnaire respondents should be followed up with interviews to clarify answers, maintain an open dialogue, verify requirements, and identify precise problems. More specifically, subjects to be covered in a principal's interview include:

1. Job responsibilities—understanding of basic job responsibilities, problems, and needs.

2. Information and communications needs—from whom, to whom, from where; frequencies and time considerations.

3. Document creation and distribution—method of origination: shorthand, longhand, machine dictation; method of distribution: mail, TWX®, facsimile, messengers, and others.

4. Work at home—where, how long, how often; information needs; equipment needs; communications needs.

5. Principal/secretary interaction—obtain an understanding of the principal/secretary team: what it is, how it works, and how it might be improved.

6. Telephone and messages—practices, problems, usage, and needs.

7. Scheduling and calendars—procedures, problems, and needs (local, remote).

8. Filing—use and size of general files, type of files kept in office, desk, age of files, sharing of files, time required to file and access data.

9. Calculations—what, how, how much, why, and where.

10. Travel—remote and portable data entry, communications needs, and so forth.

11. Future solutions—costs, benefits, constraints, and risks.

Topics to be covered in a typical secretarial or clerical interview should include items such as:

Problems
Typing
Proofreading
Confirmation of receipt
Calculations
Calendar

Lists and logs

Messages

Searching for information

Work at home

Use of terminal

Mail handling

Copying

Backup

Special items (ordering supplies, meals)

Future solutions

Obviously, many questions can be asked in any category. Questions may be open-ended or closed-ended. Some questions may be structured, others may be unstructured; some may be quantitative, others may be subjective.

In addition to questionnaires and interviews, certain tasks require the recording of data, either in self-recording logs or in logs in which the study team records its activities. Typical areas that should be studied through sampling logs include:

Support activities log (see sample in Figure 6-1)

Distribution lists

Typing

Incoming/outgoing mail—letters, reports, magazines, and other mail

Copier usage/graphics

File document utilization

Print shop

Telephone usage

Message/telecopier/other communications

Manuals (usage of policy and procedures manuals)

Any sizable office systems study should include a data analysis phase utilizing automated techniques to help in consolidating, processing, and reporting the data. The use of either an internal or an external service is optional; the choice is generally an economic one. In either case, careful attention must be given to the coding of the data entry forms and the data entry process itself, to ensure that errors are kept to a minimum.

CLASSIFICATION AND PRESENTATION

A typical office systems study will generate a great deal of quantitative and qualitative data. The proper use of the evaluation and classification procedures will require a detailed analysis of many tools. As the tools are interpreted, these suggestions should be followed:

1. Select the right sample size to ensure statistical validity.

2. Do not sacrifice effectiveness (thoroughness) for efficiency (speed) of interpretation.

3. Consider the status quo of the current environment versus the future environment.

4. Evaluate the short-run versus the long-run questions.

5. Study the conditions under which the data collection process occurred (e.g., controlled versus uncontrolled experiment, real world versus laboratory environment).[2]

James Bair (formerly of the Stanford Research Institute) discusses various analysis tools in a study conducted for the National Archives and Records Service.[3] The study provides additional insight into the use of analytical processes.

Use of Computers and Presentation of Data

A major part of the data collection activity should yield quantitative results. The system designed to interpret and consolidate data should be very flexible and generally parameter driven. Examples of some of the typical reports and statistics that should be made available by the system are: number of principals by job levels,

SUPPORT ACTIVITIES LOG   NAME _____   DATE _____

1	2 3 4	5 6 7	8	9	0 0	11	12 13

| Administrative | | | | Copying | Dictation (taking short-hand) | Errands (deliveries, pick-ups, special errands, etc.) | Information Storage & Retrieval | | | Mail | Phone (excluding meeting arrangements) | Proof-reading/editing | Typing | Other No. 1 | Other No. 2 |
Manual record keeping	Para-professional, clerical	Meeting arrangements & calendar management	Other administrative				Filing (categorizing, placing in files, etc.)	Retrieval of documented information	Informing, explaining, reception						
16 17 18	19 20 21	22 23 24	25 26 27	28 29 30	31 32 33	34 35 36	37 38 39	40 41 42	43 44 45	46 47 48	49 50 51	52 53 54	55 56 57	58 59 60	61 62 63

ADMINISTRATIVE:

- MANUAL RECORD KEEPING
  - attendance records, expense accounts, address lists, budgets, maintaining "tickler" or suspense systems, etc.

- PARAPROFESSIONAL/CLERICAL ASSISTANT:
  - Composing, drafting, correspondence, calculating, order processing, research, analysis, etc.

- OTHER ADMINISTRATIVE:
  - Travel arrangements, data input, use of "automated" reports and data bases, assembling/arranging for publication of reports, manuals and other formal documents, etc.

9	1	0
11	12	13

Number of Items ||||		Number of Telephone Calls ||||			
Filed	Retrieved	ANSWERED		Origi-nated
		No Action	Action Taken	
16 / 17	18 / 19	20 / 21	22 / 23	24 / 25

— TASK FREQUENCIES —

Copying	Errands	Filing	Retrieving							
26 / 27	28 / 29	30 / 31	32	33	34	35	36	37	38	39	40	41

Figure 6-1 Support Activities Log.

number of principals by department, and ratio of secretaries to principals and other staff members. Depending on the specific functions or areas being analyzed, additional statistics on typing, mail volumes, filing volumes and space requirements, communications, and time by job category may be required. Examples of computer and manual output forms and reports are provided in Figure 6-2 and in Tables 6-3, 6-4, and 6-5. These contain a variety of statistics covering a wide range of office activities, such as the number and distribution of telephone observations in a department, the percentage of time spent by principals at various activities, and the economics of documented administrative communications. The three examples also demonstrate that the large number of statistics that can be obtained must be carefully analyzed for meaningful conclusions and recommendations.

OFFICE SYSTEMS DEVELOPMENT METHODOLOGY

The next step in defining office systems needs and requirements and in matching the systems technologies to support these needs involves a structured approach to office systems development. A variety of approaches have been developed by IBM, N. Weinberg, E. Yourdon, A. D. Little, Inc. and others. A brief summary of the

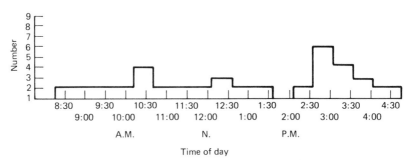

Figure 6-2   Distribution of telephone observations in a department (MIS).

Table 6-3 Principal Activities by Job Level

Activity	Percent of Time per Week[a]				
	1	2	3	4	5
Writing	4.9	8.2	13.4	14.1	12.3
Dictating	8.7	3.8	2.0	0.4	0.8
Proofreading	2.5	1.3	2.4	2.0	4.3
Reading	20.7	11.9	11.3	8.4	9.5
Incoming mail handling	2.5	3.9	4.0	6.1	3.8
File handling	1.6	0.6	1.3	2.3	3.5
Other Paper handling	1.1	0.7	0.8	1.5	2.1
Searching for information	1.0	2.2	3.5	5.6	6.4
Copying	0.2	0.6	1.0	1.8	2.0
Calculating	3.9	2.6	4.2	5.6	13.2
Scheduling	0.6	0.8	2.4	3.1	2.5
Telephoning	9.8	7.2	10.2	13.1	11.0
Typing	1.1	0.2	0.3	2.1	1.9
Meetings	38.2	47.1	37.5	28.1	20.6
Travel	2.1	3.1	2.0	1.0	0.9
Waiting	1.1	1.7	1.3	1.5	2.0
Other	—	4.1	2.4	3.3	3.2
	100.0%	100.0%	100.0%	100.0%	100.0%
Total hours[b]	43.0	44.2	43.5	37.5	38.5

[a]Column designations are: 1 = executive; 2 = senior management; 3 = middle management; 4 = first-line management; 5 = professionals; based on a 15% sample size.
[b]Excludes lunch time.

117

Table 6-4

Activity	Percent of Time
Typing	20.2
Dictation	4.6
Telephoning	5.6
Scheduling	1.3
Follow-up	1.2
Mail	3.5
Filing	4.2
Copying	10.7
Coffee break	3.2
Lunch	12.4
Statistical work	1.2
Distribution	4.0
Reception and errands	5.2
Personal work	0.8
Record keeping	1.2
Special projects	6.0
Supplies	0.6
Other	14.1
	100.0%

[a]Sample = 20 secretaries; 1000 observations, 2-week period.

Table 6-5   Filed Document Sampling at a Typical Corporate Headquarters

Document Description	Percent of File Documents
Type	
Letter	15.8%
Memo	15.2
Computer report	2.1
Typed report	13.7
Form	17.7
Teletype, Telex	1.8

**Table 6-5   File Document Sampling at a
Typical Corporate Headquarters**

Document Description	Percent of File Documents
Procedures bulletins, manuals	5.9
Annual report	1.1
Magazine	1.0
Book	0.7
Newspaper	0.4
Article, clipping	2.4
Brochure	3.7
List	4.4
Work notes	1.8
Press release	0.2
Other	11.9
Total	100.0%
Age (In Months)	
Mean	39.1
Median	23
Number Of Pages	
Mean	9.1
Median	1
Size	
8½ × 5½	1.3
8½ × 11	87.4
8½ × 14	2.9
11 × 14	1.0
Other	7.4
Total	100.0%
Kind of Copy	
Original	16.1
Machine Copy	45.9
Carbon Copy	6.8
Other	31.2
Total	100.0%

typical sequence of an office systems development methodology (OSDM) follows:

Project initiation. Make a quick assessment of the size of the problem or potential opportunity and determine whether further study is appropriate.

Initial survey. To develop a work plan, study the user's business functions, organization, plans, and information needs to clarify scope, objectives, constraints, alternatives, and solutions.

Feasibility study. Evaluate economic, operational development, and technical feasibility of alternative solutions and recommend a solution.

In-depth study. In detail, study the current system and define the functional requirements for the new or revised system or service.

Business system design. Develop and provide a business solution (a functional description of the new system and technology and how it would work) that satisfies the functional requirements stated in the in-depth study phase.

Computer (technical) system design. Specify an office system that satisfies the requirements; develop detail specifications, test criteria, and hardware/software/environmental requirements; identify operational requirements and program development phases.

Program design and programming. Design, code, and test programs (where applicable).

Conversion. Design and coordinate the implementation plan for conversion from the current system or service to the new system or service.

User procedures and training. Design forms, write procedures, and conduct training to prepare the user environment for installation and operations.

Implementation. Assist in the installation and testing of user acceptance of the new system or service, including full user and technical documentation, procedures, facilities, and other matters.

Post-installation audit. Evaluate the operational system against original objectives, user satisfaction, and cost/benefit estimates.

Each step in the OSDM should be reviewed and approved by management. The range of automation possibilities identified and presented in this manner is endless. The variables that will impact the choices include (1) type of business, (2) geography, (3) application complexity, (4) type of function to be supported, (5) organizational level being supported, and (6) organizational scope and unit. Chapter 7 describes how to match requirements with technology.

PITFALLS TO BE AVOIDED

Many office systems technologies are being developed at a rapid pace. At the same time, manufacturers and vendors of these products are claiming sizable increases in productivity. In many cases, however, office systems have been mismatched with user needs and have therefore not yielded anywhere near the actual or expected productivity gains that should have been achieved. Some of the reasons why these gains have not been achieved are presented below.

1. Sufficient analyses have not been made of the requirements and associated benefits. A compromise must be found between studying a proposed process to the most minute detail and plunging into it with insufficient knowledge or preparation.
2. Planners, users, and managers do not devote enough time to obtain a sufficient understanding of office activities and of the decision support processes that typically occur in an office.
3. Insufficient attention has been paid to the need for integrating office processes with technology.
4. Planners and analysts have not consulted or communicated adequately with the users of the technology.

KEY IDEAS

Many principles developed and applied in analyzing factory and blue-collar work can also be applied in the office environment. These include such techniques as work breakdown structures, work flows, and time and motion studies. The key to proper office systems work flow analysis is the identification of the activities, processes, and procedures followed in all offices. The basic study methodology presented consists of the following steps:

1. Establish the study objectives.
2. Identify the study team.
3. Identify the scope of the study effort.
4. Develop the data collection methodology, value analyses, and validation tools and techniques.
5. Conduct prestudy orientation and communication sessions with the participants.
6. Conduct, classify, and analyze the study results.
7. Present findings to management.
8. Prepare and implement action programs.

Data collection consists of a combination of tools, such as structured and unstructured interviews, questionnaires and observations, self-recorded logs, study team recorded logs, and automatically re-corded system usage meters. Sample size selection and data analysis must include sound validation methods. Computers can be benefi-cial in the data analysis phases of consolidating, averaging, and reporting. Once the requirements have been identified, the final phase in matching the requirements with the technology requires the use of a process known as office systems development method-ology. Depending on the requirements, numerous choices are avail-able. The ultimate decision should obviously be based on satisfying a critical need that is cost justifiable and provides a contribution, either directly or indirectly, to the bottom line.

REFERENCES

1. See, for example: F. W. Taylor, *Shop Management* (New York: Harper & Row, 1911); F. B. Gilbreth, *Motion Study* (New York: Van Nostrand Reinhold, 1911); R. L. Morrow, *Motion Economy and Work Measurement* (New York: Ronald, 1957); L. P. Alford, and J. R. Barges, *Production Handbook* (New York: Ronald, 1950);

E. S. Buffa, *Modern Production Management* (New York: Wiley, 1969); E. V. Krick *Methods Engineering* (New York: Wiley, 1962); H. P. Cremach, *Work Study in the Office* (London, England: Macalaren & Sons, 1969); Demms A. Whitmore, *Measurement and Control of Indirect Work* (New York: American Elsevier, 1979); Brian Champness, *The Measurement and Prediction of Acceptability* (London, England: Communications Studies Group, University College, 1971).

2. David W. Conrath, "Measuring the Impact of Office Automation Technology Needs, Methods and Consequences," *Proceedings of Office Automation Conference:* Carmel, CA; Stanford University, 1980.

3. James H. Bair, *Productivity Assessment of Office Automation Systems*, Vols. I and II, Menlo Park, CA: SRI International. (Prepared for National Archive and Records Service, Office Records Management, Washington, D. C., March 1979.)

CHAPTER SEVEN

What Are the Requirements and Applications?

Chapter 6 presented a study methodology for collecting data and developing findings. This chapter establishes a framework for analysis of the data gathered, as the next logical step in the development and identification of opportunities.

Once the survey has been completed, the detailed documentation of the functions, levels, and supporting processes will provide the analyst with a frame of reference for determining the applicability of technology in support of the business system components presented in Chapter 1.

Framework for Analysis

Business System Components[1]

The data must now be analyzed in terms of the following major business system components as defined in Chapter 1:

Creation

Capture

Keyboarding

Distribution

Expansion

Storage and retrieval

Disposal

The reason for this breakdown is to segment the data into meaningful categories representative of the office. Work in the office consists of various percentages of time spent on these activities. By focusing on this segmentation, the planner will gain an understanding of how employees spend their workday (see Table 7-1).

Job Classification (Employee Levels)

The organizational level of professional/managerial workers is a prime determinant of the extent of their responsibilities and thus results in the assignment of specific work activities. Because such

126

Table 7-1 Activities by Job Level (Middle
Management)

Activity[a]	Percent of Time Spent per Week
Writing	13.4
Dictating	2.0
Proofreading	2.4
Reading	11.3
Incoming-mail handling	4.0
File handling	1.3
Other paper handling	0.8
Searching for information	3.5
Copying	1.0
Calculating	4.2
Scheduling	2.4
Telephoning	10.2
Typing	0.3
Meetings	37.5
Travel	2.0
Waiting	1.3
Other	2.4
	100.0%

[a]15% sample; excludes lunch.

activities vary by job level, how and why a professional or manager interacts with office systems technology will be determined (at least partially) by employment level.

The primary reason for reviewing job classification levels is the variation of the responsibilities at each level. For example, if we consider the creation component of time spent, the potential for improvement and the functional applicability could be illustrated as shown in Table 7-2. Job classification levels will vary among organizations, but the following classifications will serve as a general reference:

Table 7-2   Job Classification-Office Systems Potential—Creation Time

Job Classification	Overall Potential	Function
Executive	Low	Speeches Presentations Reports
Senior management	Modest	Reports Schedules Correspondence
Middle management	Good	Reports Orders Decisions
First-line management	Good	Reports Orders Decisions
Professional staff	High	Reports

Management—Job Classifications:

  Chief executive officer

  Vice presidents

  Upper management (general managers and directors)

  Middle management (managers)

  First-line management (supervisors)

Professional and Technical—Job Classifications:

  Financial analyst

  Systems analyst

  Market researcher

  Engineer

  Programmer

Secretarial and Clerical—Job Classifications:

  Executive secretary (administrative and nonadministrative)

  Management secretary (general manager and director)

Secretary (middle and first-line management)

Office worker

Clerk/typist

Other nonexempt staff

The amount of secretarial and clerical support a principal receives is directly related to the job level. Exempt staff at higher levels typically receive more support than managers and non-exempt staff at lower levels. Current thinking assumes that, with office systems, the amount and quality of administrative support for principals will improve at all job levels; however, it is likely that at higher management levels, secretarial and clerical support staffs will continue to be the direct interface with office systems.

Because of the scarcity of secretarial and clerical support staffs available to professional and managerial personnel today, these groups will be receptive to direct use of office systems to get their jobs done. Despite this, it will be a long time before the widespread proliferation of management "work stations" into the executive office becomes a common phenomenon.

Functional Needs

What then are the key reasons for developing a functional analysis of the office? According to Michael Hammer and Marvin Sirbu,[2] some of the reasons include:

1. Understanding the business.
2. Avoiding the trap of becoming technology-driven.
3. Being positioned to measure value in business terms instead of in technology terms.

The focus is on two primary functional areas: interfunctional requirements and intrafunctional requirements.

Interfunctional Requirements

Because of interaction, interdependability, and interrelationships among various departments and functions, such as operating units with corporate finance or corporate personnel, it is necessary to formulate requirements on a corporate basis. For example, corporate-wide data bases, communication networks, and similar applications are considered as interfunctional (key to integration) needs.

Intrafunctional Requirements

Intrafunctional requirements relate directly to the particular needs of a specific department. For instance, these include localized text processing or duplication needs that do not have corporate-wide implications. The nature and scope of a department's functions and responsibilities determine its specific activities and communication patterns. This factor will strongly impact how principals relate to and use office systems. For example, most marketing departments require flexible communication systems to communicate to various local and distributed corporate functions as well as to external clients and suppliers.

Automation Potential

Table 7-3 illustrates a sample method of presenting the results of the analysis by job classification within the framework of business system component activities. The table also provides a functional reference by job level, and rates the automation potential of the individual components of the business communication system by job classification.

APPLICATION OF TECHNOLOGY

To facilitate the application of technology in the identification of automation opportunities, the following list ties the business communication system to the more relevant available technologies. The technologies may support multiple activities but, in order to simplify this illustration, they have been categorized by individual components.

Creation:
  Teleconferencing
  Multifunction work stations

Calculation capability

Calendars, schedulers, diaries

Library/external data base research

Capture:

Dictation

Word processing

Optical character recognition

Source document automation

Digital voice

Computer

Keyboarding:

Typewriters

Intelligent (electronic) typewriters

Word processors (local and centralized)

Distribution:

TWX®/telex

Communicating word processors

Computer to computer

Switching or electronic mail systems (voice, data, image, text, message)

Expansion:

Micrographics

Typesetting

Graphics

Intelligent copiers

Duplication systems

Storage and retrieval:

Electronic storage/retrieval systems

Automated micrographic systems

Disposal:

Shredders

Electronic storage systems

Table 7-3 Summary of Functional Matrices

Job Type	Overall Potential	Creation		Processing (Change)		Distribution (Dissemination)		Retention (Search, Retrieval, Disposal)	
		Potential	Function	Potential	Function	Potential	Function	Potential	Function
Chief executive officer (CEO)	Low	Low	Speeches Presentations Decisions	Low	—	Low	—	Low	—
Vice president (VP)	Modest	Low	Reports Correspondence Decisions	Modest	—	Modest	—	Modest	—
Upper management (General manager and director)	Moderate	Moderate	Reports Schedules Correspondence Decisions	Moderate	Productivity data Cost data Budgets Forecasts/plans What-if analysis	Moderate	Reports Strategy Plans Orders	Good	Data Various documents

									Data
Middle Management (Director and manager)	Good	Moderate	Reports Orders Decisions	Good	Productivity data Budgets Costs Status reports Correspondence What-ifs	Good	Reports Orders	Good	Various documents
Professional Paraprofessional Technical	High	Good	Reports	Good	Various data	High	Reports	High	Documents
Executive secretary (administrative and nonadministrative)	Moderate	Good	Schedules Reports Correspondence Telephone	Moderate	Schedules Correspondence Mail Telephones	Good	Various	Moderate	Documents Reports

Table 7-3 *(Continued)*

Job Type	Overall Potential	Creation		Processing (Change)		Distribution (Dissemination)		Retention (Search, Retrieval, Disposal)	
		Potential	Function	Potential	Function	Potential	Function	Potential	Function
Management secretary (general manager and director)	High	Good	Schedules Reports Correspondence Telephone Dictation	High	Schedules Budgets Correspondence Reports Mail Telephone	Good	Various	High	Reports Data
Administrative secretary (Director and manager)	High	Good	Dictation Reports Verbal	High	Reports Schedules Mail Telephone Orders	High	Various	High	Documents Reports
Office worker	High	Good	Reports Correspondence Verbal	Good	Data Reports	Good	Various	High	Files
Clerk/typist	Good	Good	Dictation Handwritten reports	Good	Various	Good	Various	Low	Files

134

Obviously, the responsibilities of the office systems planner are complex and challenging. Technology is but one part of the formula. It is more important that the planner thoroughly understand the type of work being analyzed and the impact technology will have on people and on the work processes, so that the appropriate levels of support can be established.

Office systems is an integrated collection of components that support the operations and decision-making processes of an office. Typically, office systems will provide a facility by which office workers will work more effectively, will be better equipped to assist in making decisions, will support the storage and retrieval of information, and will be linked to the rest of the organization and the outside world. In other words, an office information system provides an environment and context in which workers can do substantive and meaningful work.[3] The following section outlines several potential applications of office systems technology relative to various application needs in three functional departments:

Finance

Marketing

Human Resources

These three departments are representative of the potential for implementing systems technology and have been selected to provide a frame of reference for systems direction and planning for some typical department requirements.

Finance

Primary Technology Needs: Computation, Text Processing, Optical Character Recognition, Communications, Duplication, Graphics and Micrographics, Inquiry, and Retrieval

The clerical and administrative support requirements of the finance department will create a significantly greater than usual need for the application of office systems solutions, because of the routinized applications that exist within the function. A primary requirement of the finance department is the need to access and retrieve information initiated by various departments throughout the organization.

As an example, consider the process associated with approval and payment of an invoice. When the invoice is received in the finance department, clerical support staff could access relevant information stored on the local office system to verify the existence of an approved purchase agreement, thereby confirming that the item in question had been properly ordered. Additional review of receiving documentation, also stored on the system, would ensure that the item had been received. Comparison of the approved purchase order with commitment authority procedures stored in the system would ensure adherence to established policy and procedure. For any items that were found not to be listed or that lacked the appropriate authorization approvals, the finance staff could initiate a message, via the document distribution capability of the system, to the department or individual listed on the invoice. Through this messaging facility, the staff could request the additional information or documentation or a detailed explanation relative to the item in question. Through the use of a multifont optical character reader (OCR), the invoice could be scanned into the local office system and then routed to the appropriate department for review, for attachment of additional information, or for storage for historical purposes. Use of the mathematical capability of the system would provide a quick determination of the accuracy of the billing statement in terms of taxes applied, transportation expense, national account discounts, and summary charges. For distribution and retention purposes, appropriate procedures would address storage and retrieval requirements, including the use of media such as hard disks, diskettes, source and computer-output microfilm, as well as paper storage needs.

To accommodate profit planning and related projection requirements, the ability of the office system to produce relatively simple graphic representations such as bar charts is highly desirable. This capability is particularly valuable in variance reporting against profit planning, or similar management reports. As these initial efforts are accepted and used by management, more sophisticated graphic representations will be required and should be planned, including the use of color displays.

Secondary Technology Needs: Micrographics, Filing, Retrieval, and Retention of Information

Proper filing and retrieval procedures are an intrinsic part of the finance department's responsibility. It is therefore important to the finance department that any installed office system have sophisticated electronic storage and retrieval capability. Although many corporate functions fulfill their filing requirements by using informal or ad hoc guidelines and systems for the retention of records, the finance function must adhere to rigid standards that conform to local, state, federal, and other regulatory requirements. In many cases, these requirements will extend to the storage of documents initiated in other corporate departments or in different physical locations. Thus, in addition to electronic filming and storage requirements, the use and knowledge of micrographic techniques and associated regulatory acceptance standards are vitally important. For this reason, the office systems planning effort in the finance area should include a thorough knowledge of all aspects of the use and storage of both source document and computer-output microfilm standards and techniques.

Most regulatory agencies (Internal Revenue Service, for example) have published detailed standards of acceptance for the use of micrographics. Typically, the finance department is required to furnish supporting and source information requested by local, state, federal, and other regulatory agencies during external agency audits or similar reviews. It is therefore critical that any office system utilized for storage and retrieval purposes have established procedures and guidelines regarding all stored information and that appropriate security and audit trail techniques and issues be carefully defined and established concerning access to the data. Moreover, information that exceeds retention guidelines must be purged at appropriate intervals. Particularly in the financial function it is important that the entire subject of storage, retrieval, retention, and disposal of information be carefully reviewed on a periodic basis so that established regulatory guidelines and requirements are adhered to and fulfilled.

Marketing

Primary Technology Needs: Text and Image Processing, Use of Mixed Type Styles, Column Formatting Flexibility

The marketing department has specific requirements associated with the development of advertising copy and the processes associated with promotional documents. The typical marketing promotion staff consists of copywriters, editors, art and type directors, and other specialized creative functions. Typically, the editorial staff of such a department uses typewriters for the preparation of the copy, which is usually edited several times before final approval. In each of the editing stages the copy is usually retyped before it is presented for final approval. Moreover, it is not until late in the document production process that the editorial staff has an opportunity to review the selected type styles and page formats chosen by the art designer or type director.

Because of the involvement of professional and managerial staff in the editorial process, it is important that the office systems selected have an advanced and easy-to-use text editor, suitable for accommodating the editing and revision requirements of this function. The system should also have additional advanced text processing capabilities, such as function keys instead of code-oriented function software, and a capability for dual-page display as well as flexible column formatting, for the development of speeches and special scripts. It is also important that this department have the flexibility to use mixed type styles in a single document. For example, a speech may be printed in one font in one size on the left side of the document, and another font in a larger size on the right side. Use of these varied fonts and sizes will enable the speaker to easily read the text while essentially stating the highlights or other key statements not included in the distributed copy of the speech but critical to the theme of the talk.

Secondary Technology Needs: Type and Size Manipulation, Image Input

A number of recently introduced office systems offer type style and size manipulation through either keyboard or touch screen input. Thus the editorial staff is able to visualize the type styles as they will appear in the final copy. The more advanced systems also permit the user to establish or modify type size as well as type style. In certain systems, images (photography, drawings, etc.) can also be scanned into the office system and subjected to the same manipulation as type styles and type sizes. A number of these systems have been installed in newspaper, catalog, and magazine production facilities. The systems seem suitable for the marketing office as they become less expensive. They offer easy-to-use instructions and user transparency, and they meet office facility requirements regarding electrical, space, and noise considerations.

Use of these systems will provide the marketing staff with an opportunity to interact creatively with promotional, editorial, advertising, or other similar documents. Thus the editorial staff will not only have the ability to create the original text, but the flexibility to participate with art designers and type directors in the creative process that develops the final document. These advantages, in conjunction with the manipulation of images that have been entered as input and stored on the system (perhaps in color), will prove to be a creative and invaluable tool for the marketing staffs concerned with typography and artistic copy requirements.

Several systems that provide the capability to optically scan photographs and digitize the scanned image (including color characteristics) have been developed and offered commercially. The image or multiple images may be called up on a screen for review and modification as required. In addition to sizing and other alterations, the image's color characteristics may be modified—for example, a model's brown eye color can be changed to green, or a red dress can be changed to white. As technology progresses, systems incorporating multiple-image, text, and scanning capabilities will contribute significantly to the copy, graphic, page makeup, and printing processes typically found in the marketing department of most organizations.

As a final point, because of the unique production requirements of the marketing function, special attention must be paid to the production of high-quality, high-speed print output. This requirement will demand various technologies, including intelligent laser printers, phototypesetters, and photocomposers.

Human Resources

Technology Needs: Printing, Optical Scanning, Records/List Processing and Electronic Distribution

The human resources department's use of an office system may be more structured than that of other functional areas. This function will use the records processing capability of an office system, with specific pieces of information extracted from multiple documents. Human resources generally must review and distribute documents received from external sources. This department is also required to print stored acceptance, and rejection, or other correspondence with variable text insertions. Most personnel departments receive hundreds of résumés and applications each week. One of the primary applications of an office system in the human resources department would be to handle these applications swiftly and accurately, along guidelines carefully established in the system. As résumés and applications are submitted, the secretarial staff could identify and capture specific items of information for multiple needs, such as regulatory reporting, either by keying or by using an OCR scanner. At regular intervals (daily, weekly, or monthly, depending on the need), reports could be produced using the list and records processing functions of the office system, and these could be distributed to internal departments or to external agencies, as required.

Employment forms and related documentation (résumés, etc.) could be electronically entered into the office system by the administrative staff of the department and forwarded to the professional personnel staff, who could read and redistribute suitable applications to other personnel staff. In the process of handling the applications, the personnel staff could electronically distribute résumés or applications to managers in individual departments with specific open employment requisitions. Before an applicant is hired, additional appropriate reporting information could be entered into the office system by the secretarial staff. In handling applications and résumés, the personnel clerical staff could select master letters of various types from a stored file, print them with variable textual insertions, and mail them to applicants, as appropriate. The system could segregate and extract variable information from information previously entered by the secretarial staff. Other variables (date of application, initial contact, date of employment, etc.) could be entered into the office system according to formalized instructions and procedures.

Other uses for automation that might be considered by the human resources department include:

Promotion tracking/letter system

Reports/affirmative action program

Compliance with government programs/regulations

Personnel policy and procedures

Newsletters

Job descriptions and job posting

Human resource management

Benefits measurement

Professional training/seminars tracking

Educational updating

Administrative budgeting

KEY IDEAS

To understand the office, it is necessary to determine who is doing what. Establishing a proper framework for analysis demonstrates graphically how the office works, and provides the office systems department with an understanding of the sequence of work in simple, easily understood terms.

Completion and detailed documentation of the results of good systems analysis will provide those charged with the office systems function with an excellent frame of reference for implementing productivity improvements. This will include an understanding of automation potential relative to the individual components of the business communications system, by job classification levels and by inter- and intrafunctional needs.

The business communications system (BCS) sequence is categorized into the following activities, which were detailed in Chapter 1:

Creation

Capture

Keyboarding

Distribution

Expansion

Storage and retrieval

Disposal

To provide the office systems planner with specific and precise functional analysis and related systems applications, detailed reference listings are provided in Appendix B. These checklists will be useful to ensure that all alternatives are fully explored and analyzed and that appropriate recommendations for office systems activities and plans are made.

REFERENCES

1. R. E. Gilbert, *The Scope of the Automated Office,* Exxon Corporation Office Systems Technology Division, Presentation Paper, AIIE Conference, New York, January 28–30, 1980.

2. Michael Hammer and Marvin Sirbu, *What Is the Automated Office?,* Presentation Paper, NCC Office Automation Conference, Massachusetts Institute of Technology, Cambridge, MA, 1980.

3. Michael Hammer, *Why Is an Office?,* Diebold Research Program Presentation, Massachusetts Institute of Technology, Cambridge, MA, October 22–24, 1979.

CHAPTER EIGHT

How to Implement?

A successful implementation program requires that all possible activities and contingencies are anticipated and that solutions are planned. (Always assume that Murphy's Law* is in full operation.) This will include such items as specifications, user acceptance, test procedures, facilities, resource requirements, documentation, education and training, and others. An acceptance and reliability program should also be in place. The office systems department staff must be available to the user community, and a reporting system for addressing user suggestions, recommendations, or complaints must also be developed. If this system replaces an older one, a parallel operation should always be considered, and regularly scheduled user and vendor status meetings should be conducted during the pre-implementation, implementation, and post-implementation periods.

A comprehensive facility planning program must also be developed as an integral part of any implementation.

Existing floor and building plans must be modified to depict furniture and equipment layouts, and cabling, communications, and power facilities. If users will be inconvenienced during the installation process, disruptions should be minimized and the users should be notified of the disruptions prior to the installation.

Implementation is an extremely detailed, complex, and technical undertaking. It should be approached carefully, methodically, and with the regular use of detailed checklists. Without attention to detail, implementation can prove fatal to both the project and the entire office systems effort.

Overall program or specific project implementation does not begin until senior management approval is received and funding requirements are established. Plans should have been in place and hardware and software systems should have been selected on the basis of user requirements. Implementation requires unique skills, planning, experience, perseverance, and a focus on detail.

*Murphy's Law states that "everything that can go wrong, will go wrong."

PROJECT MANAGEMENT

Project management requires financial and people management skills as well as knowledge of methods and procedures. The project manager should take complete responsibility for all aspects of the project, even though many activities will be dependent on vendors, consultants, and individuals from other departments. The project manager is a coordinator of activities, an allocator of resources, and a technology specialist.

Each office systems project must be scheduled, resources must be allocated, and specific responsibilities assigned. Table 8-1 presents an example of how to itemize individual tasks that must be performed, how many calendar and people months will be required during a specific time frame, and when each "deliverable" is due.

The project schedule and implementation checklist should be structured to accommodate revision and status reporting. Without clearly defined milestones (e.g., costs, time, activities), any project is likely to be delayed and overbudget.

The size of the project may require the use of scheduling and implementation tools such as project evaluation and review technique (PERT), critical path method (CPM), or GANTT chart.*

The project manager's primary concern is to achieve user satisfaction. This objective can be easily overlooked under the pressure to complete the project on time, regardless of the nature of the delay. Completion of a project on time is an ideal goal; it is far more important, however, to satisfy user requirements even if additional time and expense are needed to do so. Projects that require significant delays in time and increases in expenditure should be reevaluated with corporate and user management at the earliest opportunity.

Three primary activities must be managed during implementation: project budgets, manpower requirements, and scheduling.

---

*Readers unfamiliar with these techniques should consult any of the numerous available sources that describe them in detail.

Table 8-1 Project Schedule and Implementation Checklist

Activity	Individual Responsible	198X												Date		Current Status
		J	F	M	A	M	J	J	A	S	O	N	D	Due	Revised	
Project implementation approval																
Financial review																
Contract signing																
Facility planning																
Facility preparation																
Software development																
Hardware development																
User/technical documentation																
Training and education																
Equipment shipping date																
Equipment installation date																
Parallel operation system cutover																
System acceptance test																
Installation audit																
Post-implementation audit																

Not only must the project (or projects) be implemented successfully with regard to function and user acceptance, but it (they) must also be implemented within budget, with the allocated staff, and on schedule. For these reasons, implementation requires considerable experience and awareness of project management disciplines (e.g., budgets, scheduling, resource allocation, status reporting).

For control purposes, it is important to develop a project profile for each project to be implemented. A project profile should include the title of the project, manager's name, description of the project, objectives, "deliverables", manpower, cost justification, and savings. Figure 8-1 is an example of a project description.

SPECIFICATIONS AND STANDARDS

Once requirements have been established, specifications and standards must be developed. To many, specifications and standards imply rigidity and inflexibility. If exercised properly, however, these controls will result in a more effective system to meet user requirements. If controls are implemented intelligently and where necessary, savings in effort and expense will result.

Obviously it would be a mistake to insist that all keyboards meet a unique specification regardless of application. On the other hand, a very necessary standard that should be insisted on is a company-wide communications protocol. Users should not be obligated to accept a single product or vendor, but vendors should be required to meet established specifications. So far, most organizations have progressed haphazardly in selecting equipment for office applications. As a result, in these situations the integration of systems is a near-term impossibility. Use of a "black box," which is a specialized piece of equipment that provides compatibility between unlike systems, may overcome this problem. Another approach is to develop customized software for use in a particular system. The long-term and ideal solution would require industry-wide standards, such as communications protocols, information formatting protocols, documentation, training, software development, facilities, and security. To avoid incompatibility and duplication of effort, to reduce risk

Project Title:	Touch Screen Interface
Project Manager:	W. M. Giles
Project Description:	Experience with office systems and display-based work stations indicate that principals, even those with keyboard dexterity, do not always take complete advantage of the system because of the inconvenience (real or perceived) in using the keyboard. By simplifying the interface to the system for the user, an improvement in system usage will result and, with that, greater efficiency. A touch screen interface will permit the principal to interact with the system via fingertip touch on the screen's surface. By utilizing a menu-driven system and requesting user touch selection, the principal will find the human/machine relationship more friendly.
Objectives:	Improve the human/machine interface for principals by implementing touch screen capabilities on existing systems.
"Deliverables":	By June 30, install touch screen interfaces on 10 work stations, report on system usage, and report progress periodically.

Manpower:	Project manager	2.0 Work-months
	Systems analyst	1.5 Work-months
	Programmer	3.0 Work-months

Costs:	Salaries and fringe	$ 24,650
	Touch screens, etc.	96,475
	Travel	2,500
	Consultants	9,500
	Other and contingencies	2,000
	Total	$135,125

Cost Justification:	To be determined by project.
Net Savings:	To be determined by project.

Figure 8-1 Project profile

and expense, and to facilitate integration and implementation, existing corporate or industry standards should be selected wherever possible.

USER REVIEWS

Users should participate fully in the problem definition stage, the system planning and implementation stages, and obviously the ongoing operation stage. User noninvolvement often results in poor user support and potential failure. In many situations it even results in what may be termed implementation sabotage. To become successful, systems require well-defined applications, a thorough needs analysis, a well planned implementation program, and, most importantly, user involvement, commitment, and approval.

VENDOR SELECTION AND SUPPORT

In this era of rapid change, dependence on a single vendor may not be the most appropriate strategy. In fact, change is probably the best rationale for the multiple sourcing of any product and service. Vendors themselves will readily admit to multiple sourcing the electronic components used in their products for the same reasons. Accordingly, the number of preferred vendors should be kept to a minimum for backup, control, and economies of scale.

To facilitate the evaluation of products and vendors, vendor selection criteria should be determined. These criteria (see Appendix D) should include functionality, service support, financial strength, contract terms and conditions, product reliability, maintainability, availability, research and development activities, and adherence to industry standards.

TEST DATA AND USER ACCEPTANCE

The objective in selecting and implementing any system is to match it as closely as possible with user requirements. To satisfy this concern, users should participate in developing representative tests or benchmarks that confirm applicability and validity. Before these tests can be run, or even developed, users must first identify and itemize requirements in priority order. The features that *must* exist receive the highest priority; next should be the features that would be of additional benefit; then, those capabilities that are desirable but not necessary; and lastly, those features that may be required in the future (during the product's estimated life).

The nature of the test will vary with the application requirements. For instance, a user that requires a word processor for basic typing support would most likely want to test its ability to create, edit, file, and print documents. This may require further testing of such features as centering, underlining while typing, right justification, search and replace, and footnoting. If proportional spacing or multiple columns are not a requirement, it is not necessary to test for these features or to weigh the results of such capabilities in a significant way.

Tests should adhere to requirements, and the results of these tests should be counted heavily in the evaluation effort. Systems should never be ordered without firsthand knowledge that the functionality required is actually available and performs as expected.

The products, services, and support offered by vendors vary over time. Vendors "of choice" this year may not be good choices next year. A major problem for the planner and decision maker is related to expected support during the life of the product. If a system is being implemented with an expected life of five years, high levels of vendor support must be expected to be available for the next five years.

CONTRACTS AND NEGOTIATIONS

The office systems department will need negotiation skills and an awareness of contracts in order to meet corporate legal requirements. The department must determine discounts, service levels, account status, terms, training, documentation, schedules, acceptance, and maintenance criteria. Master contracts may be desirable to facilitate future orders, and the terms and conditions must be reviewed by the legal, purchasing, and finance departments.

Both vendors and purchasers wish to negotiate the very best contract terms available. Considering the opposing objectives of the two parties, some form of negotiation will typically take place. For this reason, it is suggested that the office systems department does considerable contractual research to secure maximum leverage during these negotiations.

The vendor will want to determine realistic purchase volumes, methods of financing, and order periods. Volume orders usually merit discounts. If these volumes never materialize, however, there is usually a pay-back procedure at the conclusion of some agreed-upon period.

The purchaser, on the other hand, should determine what other customers are receiving in financial or other concessions. The purchaser will also need to know ship-to locations and related shipping costs, as well as training, installation, and documentation costs. If there are any variations in service support, such as response time estimates or guarantees, zone charges based on distance from service centers, and holiday schedules to ensure coverage, they should be discussed and documented. Typically, a service contract will be in force for a 12-month period, whereas the equipment may be ordered for longer periods. Under this kind of arrangement, the purchaser may attempt to get a commitment on annual maintenance increases in order to better budget and determine actual return on investment.

FACILITY PLANNING

Every new system requires facility planning. This planning will range from simple to complex (e.g., from plugging in an electronic typewriter to installing a digital telephone system). Associated facility requirements may involve electrical wiring, conduit, communications cabling, furniture relocation, security, air conditioning, or even relocation of office walls (see Table 8-2).

Regardless of system requirements, it is a mistake to assume that

Table 8-2 Facility Checklist

Project Title: Project Number:

Project Manager:

Facility Planner:

Item	Date Due	Date Ordered	Date Completed

1. Electrical outlets:

_____ 100 volt, 20 amp

_____ 220 volt, _____ amp

Comments:

2. Telephone lines:

_____ Dial up—PBX

_____ Dial up—DDD

_____ Private—hard wired

Comments:

152

3. Heat and air conditioning:
   _____ BTUs
   Ventilation required? _____
   Comments:

4. Flooring
   Tile? _____
   Static-free carpet? _____
   Other:

5. Security:
   Access passes _____
   Fire and safety controls _____

6. Existing floor and building plan:
   Yes _____ No _____

Table 8-2 *(Continued)*

Project Title: Project Number:

Project Manager:

Facility Planner:

Item	Date Due	Date Ordered	Date Completed

7. New floor plan:
   Yes ——— No ———

8. Statement of changes:

9. Other comments:
   Noise level ———
   Color ———
   Floor loading capacity ———

154

facility needs will take care of themselves. By ignoring these critical requirements, the planner runs a significant risk of implementation failure.

TRAINING AND EDUCATION

Training and education should address three distinct components: preinstallation, installation, and postinstallation requirements.

A preinstallation program must be initiated to include concept presentations, equipment training, system performance characteristics, facility and cutover plans, training techniques, and schedules. Consideration of these factors will allow enough time for the users to ask questions, raise issues, and acquire a better understanding of the system. It is far better to understand and address all known concerns before actual installation, not during or after it. This approach will avoid many problems, concerns, and fears.

Many implementation programs get off to a poor start because of user unawareness. All too often, a system is selected for a user community without its direct involvement, or only with the involvement of that group's senior or technical management. Even if this is not the situation, users must be provided with meaningful and practical information prior to installation.

Although most vendor and user organizations strive to provide meaningful installation and follow-up training programs, the majority of these efforts have had limitations. The programs tend to be quite costly on a per-capita basis, and training is thus limited to a restricted number of users and to an initial training course. As equipment costs decrease and vendor profit margins decline, training responsibility is shifting from the vendors to the users.

One serious situation that exists with a majority of office systems users is that only the most basic features of their systems are being used. Too often, users do not take advantage of the more sophisticated features for which the system was purchased. The reason for this neglect is poor user education and training, with little if any post-implementation reinforcement.

A better approach (and also a more costly one) is first to provide initial training only on the basic system capabilities. A month later a follow-up course should be given on more sophisticated functions. After another month, a review session should be given, during which the user describes the use of the new system in terms of the features used. During this review it should be determined whether additional training is necessary. Thereafter, reviews and additional training should be conducted on a regular basis to optimize system use and continually to enhance users' skills. Although this approach is expensive and time-consuming and requires a dedicated staff, it nevertheless results in more effective use of the technologies acquired. Additional education support includes internal and external newsletters, seminars on the use and application of office systems, and similar sources.

TEST DATA

Before a *new* system or application is installed, it must be tested by the vendor and the office systems department. This extra effort will ensure that only reliable systems that meet user needs are installed. It is far better to experience the potential problems of new technologies in a test environment than in a user environment. The test should include a review of system functions, installation preparedness, documentation support, training material, operating procedures, and system reliability. It is often during the initial 30 to 90 days in the life of any new product that most of the difficulties are observed. Therefore, by taking the precaution of thorough testing, a great deal of user inconvenience can be avoided.

PARALLEL OPERATION (CUT-OVER) AND USER ACCEPTANCE

In most implementations, an existing system, manual or otherwise, is being replaced by a new system. A smooth transition requires careful planning. It is highly unlikely that any new system will be 100% effective the first day after it is installed.

Typically, users must be trained, initial hardware and software problems must be resolved, and new procedures must be learned. Until the user has accepted and approved the new system, the existing system and procedures must be left in place. This approach provides critical backup and a comfort level for users.

DOCUMENTATION REQUIREMENTS

Documentation requires the following components: user procedures, vendor manuals, operations manuals, security and backup procedures, systems requirements, specifications, flows, maintenance procedures, and others.

Documentation is important because it provides continuity and direction between the office systems department staff and the users. All too often, documentation is given a low priority because of the time and effort required to develop it properly. In the long run this will result in user confusion, discontinuity of operations, and duplication of effort. As long as documentation is accomplished, it will provide a measure of safety and security.

Before developing, formalizing, and distributing internal documentation, it must be reviewed and approved by the users. If a serious commitment is to be made, established companies that specialize in this field should be consulted. The key to writing user documentation is to assume the user knows nothing about the product and must be educated in all of its features. Many internal efforts fail because documents are created by people who understand the product too well but never test or review their written material with the users. Experts tend to make presentations at the expert level. Too often they have difficulty in presenting such information to laymen in simple, easy-to-understand terms.

Within the office systems department, it is important to document all proposals, pilot projects, and project findings. Copies of all nondisclosure agreements and of requests for information, pricing, or quotes should be maintained on file, along with all other vendor correspondence. Facility plans (by user location) must be drawn up and retained along with special furniture orders or modifications.

POST-IMPLEMENTATION ASSESSMENT

Subsequent to the implementation or application of a new or an updated system, the effect of that system on the productivity of its users must be evaluated. This will require a preinstallation assessment, to establish a reference base for comparison.

The ideal environment would be one in which three departments that carry out the same function (e.g., marketing administration in regional sales offices) are evaluated prior to implementation. One department is provided with new technology designed to improve productivity; the second is offered new technology and a new methodology for the process being automated (this will be the most traumatic experience for the users); the third remains unchanged. If the nature of the business or other influences do not impact these three groups, a post-implementation audit can be performed to compare the results. Although this approach is time consuming and costly, it may prove important in order to demonstrate the expected productivity increases.

Other areas that should be evaluated after the implementation of new technology are the effectiveness of the training program, vendor support and service, documentation, the potential for additional applications, and the effectiveness of the strategic plan.

To ensure the viability of the strategic plan, an evaluation should be made annually. A complaint registered in the 1970s against U.S. industry focused upon its emphasis on short-term planning and its lack of interest in the longer term. Without minimizing the necessity for developing and executing long-range strategies, it is imperative that those projects considered to be short-term (18 months or less) be carried out as effectively as possible. There are two reasons for this urgency. First, failure to carry out today's projects skillfully will surely negate any longer-range efforts. Second, management is often nearsighted and tends to approach the world on a what-have-you-done-(correctly)-for-me-lately basis. Therefore, preservation and department longevity often depend on successful (profitable) short-term efforts. Awareness of this thought process will result in an approach that can be an extremely important subset of any longer-term strategy.

KEY IDEAS

This chapter has emphasized that the best-designed system, the most economical system, or the system with the highest return on investment does not necessarily result in successful implementation. Only those projects that are carefully planned, directed, and monitored will succeed. Successful implementation requires skills such as attention to detail, use of checklists, good people skills, excellent negotiation skills, budgeting skills, scheduling skills, education, and communication.

The potential inconveniences that may result from improper implementation planning can provide users and management with initial and sometimes irreversible negative impressions. It is therefore imperative to consider and check every possible requirement and overcome all obstacles in order to achieve a smooth implementation program. The user must be provided with preimplementation orientation, schedules, costs, facility requirements, and potential production interruptions, along with warnings about any other business risks. The implementation team must obtain user approvals on the overall effort and must continually apprise the user of any changes as they occur.

Success depends upon user acceptance and subsequent usage. Regardless of the strengths a particular manufacturer may have today, situations change. A vendor may have the best products and service and the lowest prices this year, but next year the same supplier's success and rapid growth may place a strain on its spare-parts inventory or the responsiveness of its maintenance personnel. Vendor strengths and limitations change over time. Never put "all your eggs in one basket."

Part Four

FOCUS ON
THE FUTURE

CHAPTER NINE

What Are the Technologies?

Four information forms—voice, image, text, and data—have always existed in every organization. Cost-effective integration of these information forms and supporting technologies will determine the office systems to be implemented. As these technologies continue to evolve and mature, integration will become more and more economical.

An ingredient critical to this integration is a business communications network that allows the transfer of all information forms throughout and between different organizations around the world. This business communications network will serve as the heart of any electronic office system.

This chapter describes numerous office systems technologies of which the reader must be aware, in order to make sound planning, funding, and implementation decisions. The various technologies were presented and highlighted in the sequence described in Figure 1-1. Table 9-1 illustrates these technologies and presents a projected time frame for their probable cost-effective availability.

CAPTURE AND PREPARATION

Word Processor Systems

The family of word processing products can be broken down into four major categories:

Keyboard/printer
Stand-alone display
Shared system
Mainframe, computer-based

Keyboard/Printer

The keyboard/printer was the first word processor system and for all intents and purposes refers back to the IBM Magnetic Tape Selectric® Typewriter (MTST), introduced in 1964. Many newer keyboard/printer systems include a one-line display to simplify editing, but the basic product is still a keyboard/printer device.

164

New products in this category have sophisticated microprocessor-based software for substantial editing; their prices are close to the cost of an electric typewriter. Many of these devices are available with an auxiliary memory on magnetic cards, tape, or mini-floppy diskettes. They are also available with communications support for electronically sending and receiving documents.

The keyboard/printer should be considered for very low volume typing applications where a secretary spends relatively little time at the keyboard and where the filing requirements do not have to be sophisticated or substantial.

Stand-Alone Display

A display-based system provides the user with better text editing, more on-line storage, and a faster and easier operation than keyboard/printers are able to offer. The stand-alone display also allows greater productivity by sometimes providing simultaneous keyboarding and printing. The typist is therefore able to input one document while printing out another. Initial product offerings incorporated hard-wired logic (electronics without software), thus making it very expensive to introduce new features.

Later systems were microprocessor-based, thus enabling the vendor to easily upgrade the system through new software releases. Some vendors chose to have software loaded from read-only memory (ROM). In order to update the software, a service representative was needed to make hardware changes. A more recent design approach uses random access memory (RAM), which permits software to be loaded from a diskette or via communications, thus eliminating the need for a visit by a field service representative (see Figure 9-1).

Suppliers of stand-alone display systems contributed to the concept of shared systems. By developing stand-alone printers with diskette readers, a word-processing center could use many display stations for inputting and editing, with one printer supporting the entire operation instead of one printer for each display. This approach was more economical, because one typist could rarely keep a printer continually busy.

Table 9-1 Office Systems Technologies

Function	1981	1983	1985	1987	1989	1991	1993
Capture/preparation		Intelligent typewriter					
		Digital voice/dictation			Full speech recognition input		
		Integrated text and data					
		Integrated multi-function work stations					
		Touch screens					
Storage and retrieval		Low cost mass magnetic storage					
		Electronic filing/archiving					
		Bubble memories					
		Laser/video disk memories					
		Cache memory					
Distribution		SBS/ACS/MCI/Telenet/Tymnet networks					
		Fiber optics communications					
		Digital PBX telephone systems					
		In-building local networks					
		High speed digital facsimile					
		Integrated facsimile/copiers					

Expansion and duplication

Intelligent copiers
Integrated photocompostion
High-resolution ink jet printing
Integrated micrographics

Creation

Electronic blackboard
Fixed and full video teleconferencing
Conference room projectors

Change

Evolution of computer technologies—faster and smaller ⟶
Data base management
Artifical intelligence—natural language interfaces ⟶
User friendly interfaces

1981 1983 1985 1987 1989 1991 1993

167

Figure 9-1   Stand-alone display. (Photo courtesy of IBM, Corp.)

The display itself is discussed more fully later in this chapter; it is important to note here that the first stand-alone display manufacturers chose a full-page display. The goal was to offer the user a system that approximated the world of paper.

Shared System

The first shared logic systems were developed by companies in the data processing field. These vendors decided to enter the word-processing marketplace by enhancing and modifying hardware previously developed for data processing applications. By following this approach, a data processing vendor was able to support multiple displays for processing both data and text. Thus, several users could share resources such as storage, printers, and communication facilities, as well as the processor itself. Several vendors chose to use microprocessors instead of minicomputers, which also provided the user with shared storage, communications, and printers (see Figure 9-2).

Several manufacturers introduced shared systems as new products; others decided to upgrade data processing products with word processing software and daisy wheel printers. Thus multi-

Figure 9-2 Shared system.(Photo Courtesy of Wang Laboratories.)

application systems were developed that simultaneously supported data entry at one station, programming at a second, and word processing at a third. This is a practical and expedient approach for data processing oriented companies that wish to enter the office systems market. The key to their continued success is how well they can support the office worker—a user very different from data processing personnel such as programmers and data-entry clerks.

The shared approach has resulted in a lower unit cost per keyboard. Once a specific cut-over point is reached, a shared system typically costs less per keyboard/display than the same number of stand-alone keyboard/display systems.

Main Frame, Computer Based

There are two types of main-frame-based text processing systems. One type is based on the use of in-house computer applications such as IBM's SCRIPT®. The other type is provided by external time-sharing service bureaus.

Users access these services on an interactive and as-needed basis. In both examples the intelligence and storage reside in a mainframe computer accessed via communicating terminals. Enhancements include archiving to magnetic tape as well as interfacing to photocomposition systems for high-quality printing.

Printer and Display Technology

The word processing product families described above consist of a combination of the following components: computers, memory, printers, displays, and keyboards. A description of printer and display technology follows. Computers and memory are discussed later in this chapter, and keyboards have been omitted, although alternative input devices such as touch screens and voice recognition are presented.

Printers

Character Printer. Vendors of word processing equipment typically offer a daisy wheel printer for hard-copy output. This device has proved to be an effective enhancement over older printer technologies. The advantages of this printer are its speed (450—550 words per minute), its flexibility (tractor feed for forms control, sheet feeder for letters, wide carriage for accounting documents, or dual elements for an extended character set), its simplicity, its quietness, and its quality of print.

It is likely that vendors must manufacture their own proprietary printer, to achieve competitive pricing and ultimate financial success in the office systems marketplace. This philosophy has been adhered to by most of the major office systems manufacturers.

Ink Jet Printer. Ink jet printers form characters by spraying ink onto plain paper; they print at least twice as fast as daisy wheel printers. This flexibility in character formation may evolve to the stage where it can serve not only as a word processor printer but also as a facsimile printer and a copier.

This product is better suited for paper handling than are most other printer offerings. Through the use of paper trays, the system may access letterhead paper, plain (second-page) paper, and envelopes. This is a significant advantage in such applications as letter writing. An additional benefit is that multiple-character fonts can be used for one-pass printing at reasonable speeds.

Intelligent Copier. The intelligent copier uses xerographic and laser technology to print documents that have originated on either computers or word processors. The output of a compatible computer or word processor can be transmitted to a printer/copier, along with the copying instructions. Multiple-character fonts are supported, as well as magnetic input, and these products can interleave printing with copying.

The intelligent copier is cost effective in high-quality and high-volume situations and can be used locally and remotely (via communications). This is a rapidly evolving area that is becoming highly competitive with the traditional printing technologies.

Dot Matrix Printer. A dot matrix printer is capable of printing fully formed characters of practically any font or style, including script (signatures). At a minimum, the dot matrix printer, along with electrostatic, chain, and drum printers, is certainly cost effective for draft-quality output printing. This higher-speed printer can also serve at lower speeds with higher resolution typewriter output quality for final copy, thus providing the word processing user with multiple functions.

Given the many printer technologies available, it is advisable to establish a list of printer selection criteria prior to any systems evaluation. Table 9-2 itemizes many of the most common printer features and provides a convenient selection checklist. Most printer technologies offer numerous options and alternatives, which must be carefully analyzed with respect to the user's specific needs and requirements.

Displays

Video displays created a data terminal boom in the early 1970s as a replacement for printer-based terminals (primarily to replace tele-

Table 9-2 Printer Features Selection Checklist

	Required	Optional	Cost
Speed			
Low 10—30 cps			
Medium 35—55 cps			
High 55+ cps			
Paper			
Continuous (tractor)			
Cut sheet tray (single or dual)			
Forms (number of copies or weight)			
Envelopes			
Wide carriage			
Special sizes			
Print			
Multiple fonts			
Legal ($8\frac{1}{2} \times 14$)			
Proportional spacing			
Subscripts and superscripts			
Vertical lines			
Graphics			
Color			
Special Features			
Cutting stencils			
Twin heads			
One pass—multifont			
Transparencies			
Forms control			
Sheet feeders			
Other Concerns			
Noise levels			
Ease of operation			
Master quality			
Mean time between failures			
Heat dissipation			
Dimensions			

typewriters). The same trend has been repeated in the text processing environment.

Digital Display. There are many display technologies, including raster, light-emiting diode, plasma, and liquid crystal. Predominant among display-oriented word processing systems is the raster scan digital display, which is based on home television technology. The reason for this acceptance is its price and performance relative to other display technologies.

Characters on a digital display are formed with dots, as they are on dot matrix printers. Each character is composed of a particular combination of dots. The more dots to the square inch, the higher the quality of the character presentation. Some manufacturers store multiple-character fonts, and in some cases extended character sets, to include other alphabets and special symbols. The greatest flexibility of presentation, however, is achievable through bit-mapping techniques (see Figure 9-3).

Bit-mapped displays allow each dot on the screen to be selected individually instead of in predetermined combinations (fully formed characters). Thus any character font or style can be displayed along with a complete graphics capability. Considerable emphasis is placed on the software, as well as on the hardware, of a bit-mapped cathode ray tube (CRT).

Digital display technology advanced quite rapidly during the 1970s. Both screen resolution (more dots to the inch) and screen size became available options, along with various screen phosphors and backgrounds (such as green on green or white on black) or full color.

As new applications developed, digital display technology evolved. The first display systems to find wide acceptance were those used for the airline reservation systems. These initial systems relied on a mainframe computer and required the computer's resources for nearly all of their functionality. As users and vendors became more sophisticated, more intelligence and memory were distributed to the display, thereby releasing the central computer's resources. Thus intelligent remote display terminals evolved to sup-

Figure 9-3   Bit mapped display, Xerox-Star system.
photo courtsey of Xerox Corporation.

port local editing, data entry, data capture, data manipulation, and communications.

Users found such applications as data entry and inquiry/response much more effective and efficient on display terminals than on card punches or printer terminals. Keystrokes could be buffered, edited, highlighted, and transmitted faster and cheaper. For these reasons, digital displays have become an integral part of word processing and office systems. Table 9-3 provides a display features selection checklist.

The digital display has captured both the data processing and the word processing marketplaces. As word-processing and office technology places higher emphasis on the quality and size of the display and on better resolution, larger screen systems will initially increase the display costs. The unit costs will continually decrease with increased user acceptance and the resultant higher manufacturing volumes.

Table 9-3

	Required	Optional	Cost

Screen Length
  Small–fewer than 20
    lines
  Medium–20—50 lines
  Full page-50+ lines
Screen Width
  Small—fewer than 60
    characters
  Medium—60—80
    characters
  Wide—80—256
    characters
  Multiple full page (side
    by side)
Character Options
  Single pitch
  Multiple pitch
  Multiple fonts at one
    time
  Standard size (6 lines to
    the inch)
  Legal size (8 lines to the
    inch)
  Underlining
  Subscripts and
    superscripts
  Proportional spacing
  Right-hand justification
  Special characters
Other Features
  Limited graphics (vertical
    and horizontal lines)
  Full graphics
  Color graphics
  Screen tilt
  Contrast control
  On-screen underscoring
  On-screen highlighting
  Character blinking
  Reverse video
  Heat dissipation
  Dimensions

Table 9-3 *(Continued)*

	Required	Optional	Cost
Multiple windows on screen · Movable keyboard			

The digital display is not the only available display technology. Other video terminals have been, and are still being, developed. These include liquid crystal display (LCD), light-emitting diode (LED), and plasma. A common goal of these development efforts has been to create a flat display that occupies less desk-top space.

Liquid Crystal Display. A liquid crystal display (LCD) consists of an organic liquid compound sandwiched between two glass plates, each with a transparent film of indium oxide. The indium oxide is a conductor and is typically patterned in a series of seven-segment characters, although dot matrix formats are now available.[1] Each segment has a wire leading back to one edge of the glass plate. Once charged, the segment will illuminate. One advantage that LCDs have over light-emitting diodes is readability in bright light. Researchers are now experimenting with LCD television, color LCD, and fluorescent LCD, which would be visible at low light levels.

Light-Emitting Diode Display. Light-emitting diode (LED) technology encompasses multicolors, high speed, high reliability, and a flat panel. Each LED serves as its own light source and can be individually addressed. Miniature LEDs may be used in a flat panel for alphanumerics as well as for graphics, when placed at up to 50 LEDs to the inch.

Plasma Display. The plasma display panel is designed around three thin glass panels sandwiched together. The panel in the center has an array of holes or cells. One outer panel has horizontal, transparent electrodes and the other outer panel has vertical, transparent electrodes. By introducing voltage to the correct row and column, the electrodes become capacitively coupled to a particular cell. The sandwiched panel contains a neon-nitrogen chemical, which illuminates in the selected cell when voltage is applied to the proper row and column.

Graphic Display. Two dominant display technologies are currently being promoted. The first is the vector approach, where points are located by Cartesian (X, Y) coordinates and a connectivity relationship is determined to draw the appropriate images. Blank areas need not be determined. An electron beam then moves from coordinate to coordinate to connect the points. This approach resembles a pencil-and-paper drawing.

A second popular technology is the raster scan display. A raster display screen is divided into picture elements, or pixels. Each pixel is then assigned attributes, such as color and brightness. Each pixel requires refresh memory in order to sustain the image, as with television; thus the raster technique would require more local memory than the vector (see Figure 9-4).

As of this writing, the most widely used application areas are computer-aided design (CAD) and computer-aided manufacturing

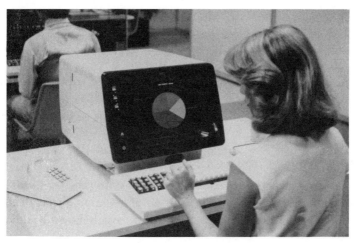

Figure 9-4 Color graphics display, IBM-3279 display terminal. photo courtesy of IBM Corporation.

(CAM). In CAD environments, the design of electronic circuits is facilitated by the graphic display or in some cases a graphics output printer. An example of a CAM application is numeric control. The parts programmer does not have to visualize and program a numerically controlled system; the CAM system can retrieve and program on the basis of stored experience factors.

The use of color graphics is expanding in offices to better represent financial and performance data. A pictorial representation of data is more easily understood than tabulated data.

Miscellaneous Technologies

Users of office systems can input both commands and information via a keyboard. Alternative methods do exist, and these will become more available and economical in the future.

Light Pen

An alternative input technique on data display terminals is the light pen. A light pen user relies on menu-oriented application software and can select from lists of options by touching the pen to the screen. Without an associated keyboard, other information cannot be entered.

Touch Screen

Another potential input mechanism is the touch-sensitive screen. This approach consists of a special display screen with vertical and horizontal (*X,Y* coordinate) crossings. The user need only touch his/her finger to the screen at an option point. The touch screen is very similar to the light-pen approach and has some of the same limitations.

Voice Recognition

Much progress has been made over the past 10 years in the area of audio technology. More specifically, audio technology has been developing in two distinct areas: speech recognition and audio response.

Speech recognition, also known as automatic speech recognition

(ASR) or voice data entry (VDE), utilizes speech pattern recognition techniques to identify a given word or phrase spoken by a specific person. Speech patterns are not easy to work with. Every voice has different patterns for different words, for the same words spoken under different circumstances (calm, stress, or illness), and for external influences, such as background noise. All these features can affect the recognition process.

People speak with contiguous words in strings to compose a statement or a sentence. Voice recognition systems are still not capable of interpreting with continuous speech. Automated systems have been limited to isolated word recognition where single words, numbers, or letters are spoken (or in some cases very short phrases, such as "Turn on the lights"). Breakthroughs made in this technology allow for the recognition of longer phrases, but not yet for continuous strings. This middle ground is being termed connected speech. Although the cost for a connected-speech system is many times higher than that for an isolated-word system, the former is a major step forward.

Until now, a voice recognition system in the office was even more ineffective (from the standpoint of cost/performance) than light pens and touch screens. Once connected speech is available, office systems users will be able to easily enter information inquiries and commands directly and conveniently.

The second area of audio technology is the audio response system. This system can be based on either the stored speech approach or the speech synthesis approach. The former utilizes prestored human voice recordings and is commonly used by the telephone companies for "wrong number" and "coin deposit" messages. The latter approach actually constructs speech from individual phonic units. Such an approach has been used quite successfully in a reading machine for the blind, which optically reads books and newspapers and converts the written words into spoken words.

Optical Character Recognition

Optical character recognition (OCR) equipment scans a document for recognizable characters or bar codes. Long used for capturing information for data processing applications, OCR is now being applied to office systems by some vendors and users.

An OCR machine will accept a piece of paper (typically a fixed size) and move it past an optical scanner. A beam of light horizontally scans the piece of paper, and the reflected light (white or dark) is converted to a digital signal. The system's logic, or software, then matches the patterns discerned by the light beam with those character or symbol patterns stored in the system. Many systems are quite limited in the number of character fonts that can be stored at one time. Some of the more sophisticated (and more expensive) systems can keep several fonts available for comparison, and other systems are sophisticated enough to easily learn additional type styles.

Other systems limitations being addressed by vendors include paper quality, line spacing, margin size, mechanical paper handling, and page registration. When reading errors do occur, the OCR system will typically flag the unreadable character for operator intervention.

SEARCH, STORAGE, AND RETRIEVAL

Primary (main) and secondary (auxiliary) memory techniques have continued to evolve into smaller, faster, and less costly technologies. Main memories evolved from vacuum tubes to transistors (core) and then to solid-state integrated circuits. Newer systems are beginning to utilize cache and bubble memory and charged coupled devices (CCD). Auxiliary memory devices are basically disk and magnetic tape, which continue to achieve higher densities, faster speeds, and smaller sizes. Newer concepts such as floppy disks, video disks, and mass storage systems are constantly advancing. Similar progress has also been made in micrographics.

Random Access Memory

Random access memory (RAM) is semiconductor memory. The dramatic manufacturing economies that have resulted from the large-scale integration (LSI) of these memory chips was a reason for this widespread acceptance. In 1973, core memory was predominant. By 1979, the cost of RAM was barely one-tenth the cost of core on a per-bit basis. During this six-year period RAM memory chip capacity had increased 64 times. Along with these significant density increases have come increases in speed. In fact, solid-state RAM systems are now being offered as an alternative to fixed-head disk systems. The user can choose between mechanical systems and faster, more reliable, and less expensive solid-state systems.

Cache Memory

Another form of RAM technology is the faster cache memory. Cache memory is based on bipolar technology, which is faster than RAM but does not support the high densities. Most RAM is based on metal-oxide-silicon (MOS) technology, which can be produced at very low cost because of the high densities.

Bubble Memory

Bubbles are tiny magnetic cylinders that are moved when charged by an electric current. These bubbles are on very thin layers of a conductive material, such as garnet. When the current is applied, magnetic fields form around small oval holes found on the garnet layers. The bubble moves around the oval hole based on the polarity of the magnetic fields.

Several vendors have begun manufacturing products with bubble memory. Costs are expected to decrease considerably in the future, and the nonvolatile nature of bubbles makes them an attractive alternative to magnetic media systems, for example, floppy disks.[2] So long as its speed remains slower than RAM, bubble memory is an unlikely replacement for RAM. The bits of information stored on the bubbles are accessed serially; RAM is accessed randomly.

Charged Coupled Devices

Charged coupled devices (CCD) are stored memories that take the form of electrically charged packets that shift on a silicon oxide chip.

In fact, the structure is based on serial shift registers. Because the CCDs are serially accessed, they are unlikely replacements for RAM. However, the low cost per bit does make CCDs competitive with bubbles, and although they are accessed serially, they are extremely fast at block-oriented processes.

Magnetic Disk

The cost per bit of disk systems has continued to drop. Only tape storage remains less expensive as a magnetic media. As with solid-state memory, the major breakthroughs have produced higher densities and faster access speeds. Newer disk systems contain as many as 12,000 bits per inch. This trend toward higher density at lower costs is expected to continue.

There are a great variety of disk systems, including cartridge disks, fixed-head disks (occasionally still referred to as drums), removable disks, and nonremovable disks. Over time these systems have become faster and more reliable and offer more memory per inch than it was thought possible to achieve for this medium.

Floppy Disk

The floppy disk, also known as a diskette system, first began as a storage source for loading programs into other systems, such as a hard disk or a communications controller. It quickly became an attractive replacement for slower, magnetic tape cassette drives on intelligent terminals and on word processors.

The first diskette was eight inches in diameter with recording capability on only one side (approximately 256,000 characters per diskette). Advancements have led to double-density and double-sided diskettes with one-million-character capacities. The floppy disk now comes in various sizes, such as the five-inch diameter minifloppy and the "silver dollar" size floppy disk, which fit nicely into the terminal keyboard and also consume less energy.

Optical/Laser Disk

The optical/laser disk (also known as the video disk) provides very high density recording with low-cost storage of data, digital voice, text, and images. Retrieval is fast and a single disk can hold 54,000 frames (or 13 billion bits) of information. The laser disk is a viable alternative to the magnetic disk and micrographics for mass-storage applications.

Mass-Storage Systems

Over the years, several magnetic tape systems have been developed to support applications with very large data bases, provided that rapid retrieval speeds are not an application requirement. Such systems can store in excess of 200 billion characters of information, with retrieval taking several seconds.

Mass-storage systems are based on the access of magnetic tape cartridges by movable access mechanisms. Once the selected cartridge is accessed, it operates sequentially like a tape reel.

Electronic File Cabinet

Data-base management now extends into the office for text storage and retrieval as well as for traditional data applications. All those files that will eventually be stored on magnetic, video, or microform medias must be accessible and secure.

Regardless of where the files are located (centrally, locally, or remotely), it will be necessary to access them easily. This will require well-designed and easy-to-use indexing schemes. Information should be accessible by subject, author, addressee, date, keyword analysis, relative value, or content.

Several systems and services already exist to facilitate the storage and retrieval of documents. The problem is that many of these services require extensive training to be used efficiently. They require better human engineering, lower cost, and greater security before they can be used in the office.

DISTRIBUTION

Office systems will succeed if they provide users with economical information management. To manage information properly, fast and reliable electronic communications are necessary. In fact, communications will serve as the central nervous system providing information paths throughout the organization.

Transmission Protocols

Systems can communicate with each other only if they have the same communication software (protocols). Currently, because a number of transmission protocols are available, it is difficult to develop inter- and intraorganizational information networks. This situation was caused primarily by the vendor's desire to "lock in" customers. The reasoning is that if a manufacturer's computer or computer application can communicate only with its own terminals, the manufacturer has created an enhanced business opportunity. The voice telephone network originated as many independent, incompatible systems. If this incompatibility had continued, customers would have been denied the flexibility and additional uses offered by the telephone today.

To add to the confusion, certain protocols are communicated asynchronously (character-oriented), synchronously (block-oriented), or under data link concepts (bit-oriented). Even if two vendors characterize their equipment as compatible with one of these protocols under a particular transmission technique, compatibility is not guaranteed for several other reasons. Communications can occur at different speeds (bits per second) and using different code sets (e.g., American Standard Code for Information Interchange or ASCII, and Extended Binary Coded Decimal Information Code or EBCDIC). There is also the issue of modem compatibility (the device that interfaces a communications system with the telephone line), and each protocol usually has several options with which a particular modem supplier may or may not offer compatibility.

Formatting Protocols

The information being transmitted is generally formatted differently by each vendor. For instance, two manufacturers that have "compatible" protocols may differ on how they format information (e.g., interpret margins, rules, underscores, or center codes).

Industry, government, and professional and trade associations are aware of these problems and are attempting to standardize

communications protocols on a domestic and international basis.

Dial-Up or Private Line

Two basic types of communication networks exist with regard to the
physical access of one piece of equipment by another. One type is
a dial-up network. In a dial-up, or switched, network, such as that
provided by AT&T for voice calls or by Western Union for TWX®
communications, the user at one location dials (or pulses) the number at the second location. The call is then routed over the network
via nodes in any of a number of different possible ways.

The other type is the private-line network. Private-line networks
are usually found in reservation systems or message switching systems, where the information is required rapidly, the information
volumes are high, or security is a major concern.

If properly engineered (conditioned), private-line networks usually permit faster transmission than do dial-up facilities. Limitations
of private-line systems include: (1) the need to reconfigure the
network whenever changes are made, (2) the need to achieve the
least possible cost routing; (3) configuration changes take longer
than they do on dial-up networks, and (4) if a segment of the network malfunctions, all systems on that segment will not operate.

IntraBuilding/Local Networks

Since the majority of communications (50%–80%) occur within a
specific facility or building complex, local area networks offer significant opportunities for integration. Although this technology
holds promise, its realization will be affected by its newness and the
lack of agreement among vendors concerning standards. Ideally,
local area networks will support voice, data, image, text, and video.
In addition, a network can have intelligence to provide compatibility
and linkages between unlike systems. This network may consist of
coaxial cable, fiber optic cable, twisted copper pairs of wires, or
combinations of all three.

Local networks will be either baseband (one communications
channel) or broadband (multiple communications channels). They
will be passive (no computer control necessary for routing) or they
may be computer controlled. Numerous options are open to the

user, and the solutions will depend on vendor standardization and cooperation, in addition to user requirements.

Fiber Optics

The introduction of fiber optic communications technology has been a breakthrough in wiring quality, cost, and bandwidth. The quality of transmission is much better than that of copper wire, principally because fiber optics is immune to electrical noise. Fiber optics is more cost effective as a result of its larger bandwidth, which reduces the cost per communications channel.[3]

The system contains three major components. The first is the light source; either light-emitting diodes or injection-laser diodes are used. So far, the problem has been the relatively short life of these light sources, especially for lasers. The second component is the cable itself. The center of the cable is cylindrical glass fiber. Surrounding this center is a layer known as the cladding. The center fiber has an index of refraction higher than its cladding, and this causes the light source to be reflected forward along the length of cable. Some of the light's intensity will be lost as a result of reflection and fiber impurities. The third component is the optical detector, which converts the light signals back into electrical signals.

Fiber optics is already being used by various telephone companies primarily for voice communications, by cable television (CATV) companies for video and voice transmission, and by equipment manufacturers for interfacing to peripherals.

TWX® and Telex®

TWX® and Telex® are Western Union teletypewriter message networks. TWX® exists only in North America; Telex® is worldwide. Both services have evolved from the earlier Telegraph® system. They utilize the privately owned Western Union dial-up communications network in the United States and interface to all other international carriers for access to other countries.

Infomaster®

Western Union provides a store-and-forward communications service called Infomaster® in order to interface and facilitate its many communications applications. Infomaster® provides an interface between the TWX® and the Telex® networks as well as many TWX® and Telex® enhancements. These additional features include distribution and re-entry capabilities. Infomaster® also provides access to Mailgram®, to facilitate access to individuals and other nonteletypewriter users.

Advanced Communications Service (ACS)

AT&T has announced and submitted to the Federal Communications Commission (FCC) plans for ACS® to provide on a nationwide basis a packet-switching network that is capable of communicating among various types of terminals and computers. ACS® will offer its customers a communications network that can interface many different terminal types, convert protocols and codes, store and forward messages, do limited input editing, and monitor usage and performance.[4]

Satellite Business Systems

It is the intention of satellite business systems (SBS) to provide broadband communications in the United States via satellite. Customers will have to install earth stations to transmit and receive via the satellite. As a result of this broadband communication path, users will be able to transmit not only voice and data traffic but image and video traffic as well. This allows for full video teleconferencing and high-speed facsimile.

Electronic Mail and Message Systems

Until recently, only traditional message-switching networks existed. Corporations then started using teletypewriter-based, private-line, and store-and-forward message switching for communications between remote locations. These systems still exist because few economical alternatives have been available.

There are alternatives, however. Users can send or pick up messages to or from other subscribers. These systems support time-sharing terminals including lightweight portable units that are carried easily. Most electronic mail networks such as these can be accessed locally via several communications networks. Each user typically has an individual "mailbox" into which all incoming messages are deposited. The messages stay in the mailbox until the addressee reads them and takes some action.

Another form of electronic mail and message system has developed in Europe on a limited basis. By using the home telephone and/or television, a person can inquire into data bases for personal or business use. For instance, train and movie schedules can be obtained, stock exchange prices can be checked, or computerized television games can be played. A person may also be able to access an on-line telephone directory, send and receive facsimile documents, or access an office system right from the living room.[5]

Store-and-Forward Voice

Store-and-forward voice is similar to store-and-forward message systems in that the former deals with speech and the latter deals with written messages.

In this system, telephones operate like sophisticated answering recorders. If the party being called does not answer the call, a recording requests the caller to leave a message. When the caller disconnects, an indicator light comes on, notifying the receiving party that a message has been received. The receiving party enters a playback code and listens to the recording, and selectively saves, replies, forwards, or discards the message. Any touch pad input telephone serves as a terminal. Addressing may be by a code (e.g., extension number) or, depending on the level of directory sophistication, by name.

Facsimile

Although facsimile technology has existed since 1842, it did not become a widely accepted business tool until the mid 1960s. Interestingly, facsimile has not been used to any significant extent for any new applications. Rather, it has proved successful because it sends and receives documents faster than existing mail services.[6]

Facsimile is similar to intelligent copier systems in that a page of information (text, data, or graphics) is scanned and an analog or digital signal is created to represent the black-on-white image on the paper. That signal is then transmitted over telephone lines to another compatible unit, and the signal is converted back for creation of a new hard copy at the receiving station. Facsimile and communicating copying are so similar that they will eventually become one and the same.

Facsimile equipment vendors have developed a range of analog and digital products offering six-minute transmission speeds with 100% operator control for feeding in documents one page at a time and for establishing the telephone connection, and 20-second transmissions with paper-feeding mechanisms and automatic dialing units. These higher speed systems may cost from five to ten times more than the operator-controlled ones. The purchaser's decision must therefore be based on usage volumes. As the number of pages to be transmitted each month increases, the use of the faster system becomes more economical. This is even more evident when the labor costs are considered.

Some of the most recent facsimile products offer compatibility with both analog and digital transmission, as well as acceptance and transmission from computer-generated output, TWX®, telex, or textual information.

Much of the facsimile traffic is textual and timely (e.g., memos and sales order data). The mail services could not compete because of the time factor. As electronic keyboards become less expensive, communications between word processing systems will replace some use of facsimile. Once a document is prepared on a word processor, it can be sent directly to its destination without having to be printed and taken to the facsimile system to be transmitted.

EXPANSION AND DUPLICATION

Micrographics

First patented in 1859, microphotography was introduced by Réné Dagron, a French photographer. This technique of reducing an image many times onto film is now widely used for many different applications, such as computer output and the widespread distribution of manuals.

Microfiche is a matrix of microfilm images on a sheet, instead of sequentially on a roll (microfilm). Reduction capabilities are extremely high, thus permitting thousands of images on a microfiche sheet. This high density is called ultrafiche.

By computerizing the directory of information in a microfiche file, it is possible to enter a request on a terminal and automatically retrieve the proper image. The next stage of development would be to read this image and convert it to a digital code that can be displayed on a bit-mapped display or transmitted to a computer.

Several companies are attempting to write the micrographic image with a machine-readable font so that an optical scanner can be used to read the film for processing. Systems can now transmit a microfiche image via facsimile to a remote location, thus making fiche a more effective storage medium.

Photocomposition

Most printed material, such as books, magazines, and brochures, are now produced on photocomposition (phototypesetting) equipment. This equipment takes keyboarded text and digitized images, combines them, and then photographs them for output. Just prior to printing, the text on a line can be justified to evenly fill the space provided (proportional spacing) and line-ending words can be hyphenated automatically.

Previously, typed information was rekeyed into the typesetting system. This duplication of typing was costly in time, money, and accuracy. Formats, logos, and other information can now be prestored with a considerable reduction in effort. The rekeying step is being eliminated as communications and media compatibility between word processors and photocomposers has become more commonplace.[7]

Newer electronic systems may eventually replace the entire photographic process. These electronic cameras can attain over 17,000 dots per square inch and may eventually be adapted to color as well. Electronic photographs can be transmitted, stored, edited, sized, and color corrected. Output can also be microfilmed.

Computer-aided page displays are now being used to make up

pages for final printing. They permit the combining of graphics and text for on-line page makeup.

CREATION

Teleconferencing

Teleconferencing is intended to reduce the travel requirement and the associated costs, while permitting greater participation at meetings by more people, via electronic devices. The various methods of teleconferencing include voice conference calls, electronic mail conference calls, full-motion video conferences, fixed-frame video conferences, and a unique variation of these known as the Electronic Blackboard®.

Voice Conference Calls

Voice conference calls exist in two forms. The first method utilizes a speaker phone at one end or both ends of the conversation. In the second approach the telephone operator is requested to arrange a multiple-point telephone conference call. Some of the newer telephone systems allow for conference calls without operator assistance.

Electronic Mail Conferences

Electronic mail networks allow for several keyboard terminals to communicate with one another interactively in a conference call mode. This is more tedious than voice conferencing, because all other parties must wait for the current speaker to key in his/her remarks. The advantage of this approach is that all participants have a written record (the minutes) of the conference.

Full-Motion Video

A full-motion video system works in the same way as closed-circuit television. All activity is captured and transmitted to the distant location. However, this technique will become economically feasible between distant locations only when satellite transmission facilities are widespread. The current cost for the bandwidth necessary to transmit full-motion video is very high.

Holiday Inns and AT&T offer full-video service, but users must go to their facilities. Although this is more convenient than traveling long distances, it is far from the ideal.

Fixed-Frame Video

Another approach to video teleconferencing is fixed-frame video. This technique also utilizes cameras, monitors, microphones, and speakers; however, the video is not full motion. A new picture is transmitted several times per minute. Thus the monitor displays an image for a number of seconds until the next frame is received. Full-frame video can be transmitted in black and white or in color, and at much lower cost than full-motion video. Thus, if meetings contain little motion, the fixed-frame method of video conferencing is sufficient.

Electronic Blackboard®

The Electronic Blackboard® consists of a pressure-sensitive black-board, microphone, and speaker at one location, and a TV monitor, microphone, and speaker at a second location. As someone, for example, an instructor, writes on the blackboard, the coordinates are picked up by pressure-sensitive electronics and transmitted to the monitor, where an image of the blackboard is displayed. The two-way voice connection permits discussion of the written material. This form of teleconferencing is inexpensive and easy to install and can be accomplished over standard dial-up telephone lines.

Voice Telephone Systems

Voice systems are evolving rapidly. The digitizing and compressing of voice so that it can be stored economically on computers has opened up some exciting new application horizons.

The private branch exchange (PBX)—sometimes known as computerized branch exchange (CBX), private automatic branch exchange (PABX), or electronic private automatic branch exchange (EPABX)—has developed into a computer system capable of not only telephone call switching, but also least-cost telephone call routing, message switching, store-and-forward voice, and other features. The newest systems use fiber optic cable connections and include distributed switches.

The PBX is becoming so sophisticated that it is often termed a supercontroller, to emphasize its potential importance in networking. By integrating a PBX with an in-building cable network, it is possible to control all communications via one network. This may prove economical, but it may also prove to be too high a risk, if a cable failure should occur.

Dictation

Although voice recording on tape for later transcription has existed for some time, significant improvements have been made in user techniques and equipment types.

Minicassettes now enable pocket-size units to be carried for use while at home, commuting, traveling on business trips, or during meetings. Some of the newer microprocessor-controlled systems allow telephone access to a recorder for use by many people and from any touch-tone telephone. Such systems offer portability, convenience, queuing, priority assignment, usage statistics, cost allocation, and the potential for significant time savings.

It is reasonable to expect that the digital PBX or similar computer system will eventually become a company-wide dictation and store-and-forward voice support system.

Conference Room Projectors

The meeting or conference room has long existed as the place for sharing information and making joint decisions. As people begin to rely on electronic systems to create, change, and review information, it will become natural for them to share visual and electronic information collectively in the conference room. To do this, large conference room projectors are needed, capable not only of showing slides and films but also of displaying selected text, data, and graphics resident in the office system.

Several projection systems are available, varying from home television projectors to sophisticated light valve systems. The input to these projectors is standard digital television signals from a display with 525 rastor lines. If an office system relies on full-page, one-line, or plasma displays or any other variation, some very expensive electronics must be developed to convert from one technique to television.

CHANGE

Computers

Over the preceding four decades, revolutionary changes have occurred in the computer industry. Hardware has changed from large, expensive, slow, and difficult-to-program vacuum-tube technology to very large scale integrated circuit technology characterized by small, high-speed, low-cost, and easy-to-program technology. Rapid and dramatic changes are expected to continue in this industry.

Mini computers

It is extremely difficult to delineate between mainframecomputers and minicomputers. Previously, one could relate to such factors as physical dimensions, the number of bits per word of memory, or processing speeds. This distinction has become much more difficult in recent years because large computers have become smaller and small computers have become larger. Processing speeds and memory sizes for minicomputers have increased substantially. Minicomputers can also communicate, drive the same peripherals as maxicomputers, and operate multiple tasks simultaneously.

Microcomputers

A microcomputer, or microprocessor, is a logic device contained on one or more chips. The first microcomputer was developed for a small-scale system capable of performing business applications locally and inexpensively. Today, microcomputers, or microprocessors, are found in watches, automobiles, games, word processors, calculators, and microwave ovens. The microcomputer has progressed from integrated circuits (chips) with only four gates (logic circuits) to medium-scale integration with hundreds of gates, to large-scale integration (LSI) with thousands of gates, to larger-scale integration (VLSI) with tens of thousands of gates, and now to very, very large scale integration (VVLSI) with millions of gates. The most significant result of this density increase is that manufacturing cost is related to chip size and not to the number of circuits on a chip. As the density increases, the cost per circuit continues to decrease.

Software

Computers and office systems have become more and more dependent on software. Software has evolved to make the hardware more efficient and the user more productive. Software-based systems provide greater flexibility. Software will make office technology usable by executives, managers, professionals, and clerks.

Software systems consist of operating systems, application programs, communications programs, utilities, data-base management systems, and user interfaces.

Distributed Processing

Users often want direct control of all the resources necessary to run the business, including the computer. This control is particularly important in the area of office systems. Users want control and maintenance of local information, security of files and documents, and confidentiality, accessibility, and more flexibility. The trend is toward migration from centralized data processing and communications with office systems to a distributed environment.

Data Base Management Systems

Data base management systems (DBMS) are designed to create, update, request, process, and report information by eliminating redundancy and minimizing the need for specialized programs by separating data from program logic. If customization is necessary, programs can be written and interfaced to the data-base system.

Office systems require user oriented data-base tools such as decision support systems, ad hoc report requirement, key word, parameter, and content-driven retrieval systems, among others. Such systems are now available.

ARTIFICIAL INTELLIGENCE

Artificial intelligence facilitates the use of office systems through the use of natural language and the ability of the system to make decisions based on previous experiences.

The use of artificial intelligence techniques will allow for natural language inquiries—for example, "How many employees were hired in January?" Users can begin to work immediately with such application interfaces. This is why input, by voice, in one's native language will become the most acceptable input form. Significant research is now going on in this area at such institutions as Massachusetts Institute of Technology, Stanford University, and Carnegie-Mellon University.

KEY IDEAS

This chapter has related the business systems components with the technologies. These technologies range from word and data processors to artificial intelligence. To deal with the office systems environment adequately, it is necessary to understand these rapidly evolving technologies and their interrelationships.

Computers are becoming faster, smaller, and less expensive. Memory systems are following the same pattern. Communications networks are becoming commonplace and provide access on a local, nationwide, and worldwide basis. High-quality displays are available today and promise to improve in the future. Voice, graphic, and video technologies are also advancing rapidly. Software and communications systems will link the different pieces together so that the integrated result is an office system that provides a high degree of utility by improving information management and business communications.

The number of functions and activities that can be automated is virtually endless. Once the intelligence is available in an easy-to-use, cost-effective work station, the office worker will realize the benefits. The technology breakthroughs discussed in this chapter are moving rapidly toward this objective.

REFERENCES

1. Tomio Wada, "Liquid Crystal Research Updates LCD Technology," *Journal of Electrical Engineering*, January 1979, pp. 38–41.

2. "Bubble Memory Devices to Play Increasing Role," *Communications News*, July 1979, p. 83.

3. John Gantz, "The Secret and Promises of Fiber Optics," *Computerworld In Depth*, pp. 1–10, October 8, 1979.

4. George Davis, "AT&T Answers 15 Questions About Its Planned Service," *Data Communications*, February 1979, pp. 41–60.

5. "Turning Telephones into Terminals," *Business Week*, October 1, 1979, pp. 86–90.

6. Wayne L. Rhodes, Jr., "Facsimile—New Life for an Old Idea," *Infosystems*, September 1979, pp. 42–52.

7. Michael Kleper, *Everything You Always Wanted to Know about In-Plant Phototypesetting in Less Than One Hour* (Bedford, MA: Compugraphic Corporation, 1978).

CHAPTER TEN

Where Do We Go from Here?

Thus far we have addressed productivity improvement opportunities that will result through the introduction and use of office systems. This chapter focuses on the future and initially addresses two areas of concern. The first involves those issues that will continue to drive the office systems effort. These include:

Energy shortages and rising expenditure levels

Educational levels

Government and regulatory reporting requirements

Shortages of skilled personnel

Changes in demographic work, life, and age patterns

Improved productivity expectations

Technology innovation

Inflation

The second area of concern involves those issues that will be affected either directly or indirectly by office systems, including:

Increased specialization

Middle and other management staff displacement

Organization stratification

Increased spans of control

Depersonalization

The office of the nineties will differ significantly from the office of today. While no one can predict the exact consequences, it is generally agreed that the next decade will witness significant alterations and dislocations in our current life-styles, societal expectations, educational requirements and institutions, demographic work patterns, and career goals. It is only through an awareness of these issues that the office systems department will be able to develop an orderly and successful strategy that optimizes the use of technology and minimizes the disruption caused by the introduction of technology on people and the business environment.

200

DEALING WITH CHANGE

Although it is difficult to predict the changes and resultant societal consequences of the problems facing business today, strategies and plans must be developed that address these anticipated changes. Actually, business and society may now be embarking on a second industrial revolution. Strategic business planning in the next decade must anticipate dislocations and must be prepared to deal with them. A willingness for dealing with change must be accepted. Obsolete business methods and practices have no place in the 1980s. Greater attention must be focused on such important factors as environmental issues, the social design of the office environment, and changing work habits. While much research and literature have been developed around organizational issues, relatively little exists concerning the effects of introducing the new office technologies into the office environment (see Figure 10-1).

What will be the concerns as the office systems effort becomes more prevalent? The following representative issues will directly result in a continuation of the office systems proliferation we have witnessed over the past several years. By no means should this listing be considered all inclusive.

Energy Shortages

Business will be concerned about the shortage and high costs of energy and related resources. In most corporations, alternatives to travel will be closely investigated, and travel will be reduced significantly. Corporate goals will be established to make maximum use of all types of energy-related resources.

Educational Levels

A factor that will affect the quality of work produced will be the quality of formal education. Studies over the past decade have pointed to a significant decline in the scholastic aptitude test (SAT) scores in both mathematical and verbal categories. In certain major urban areas the majority of high school graduates are unable to pass

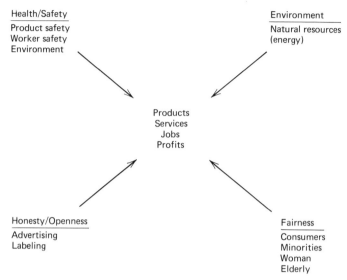

Health/Safety
Product safety
Worker safety
Environment

Environment
Natural resources
(energy)

Products
Services
Jobs
Profits

Honesty/Openness
Advertising
Labeling

Fairness
Consumers
Minorities
Woman
Elderly

Figure 10-1   New business responsibilities and pressures of the 1980s. (Adapted from
Corporate Priorities Presentation, American Council of Life Insurance, Yan-
kelovich, Skelly & White, Inc., Boston, Massachusetts.)

minimum qualifications for state diploma certification. If this trend
continues, organizations may have to pick up where the education
system leaves off. This will be especially true if the technology
emphasizes written rather than verbal skills. As Howard Morgan, of
the Wharton School of Business at the University of Pennsylvania,
has stated: "Some of the interesting observations which have
been made include the lessened effects of verbal skills, as opposed
to writing skills, among people who communicate on task-
oriented matters through electronic methods."[1] Retraining of tech-
nologically displaced personnel is another major educational requirement.

Government and Regulatory Reporting Requirements

As government continues to grow and as consumer awareness increases, we can expect a continuous increase in the already large
number of government and other regulatory reporting require-

ments. This will encourage preservation of more historical business records and will significantly increase the need for improved records management programs and related technologies such as computer-output microfilm, source document microfilm, and electronic storage techniques. We can expect the newer technologies to offer some intriguing storage, manipulation, and retrieval capabilities, which should go a long way toward resolving the technical needs associated with the increase in reporting requirements.

Shortages of Skilled Personnel

We will witness, in the next decade, a large population aggregation in the age group between 25 and 44, while concurrently experiencing a sharp decrease in available personnel with clerical, secretarial, and administrative skills. Much of this will be associated with the continued emergence of a population consisting of two wage earners in one household. A desire for upward mobility will result in rejection of lower level clerical, secretarial, and administrative job opportunities and a severe shortage of available personnel to fill these openings.

Demographic Work, Life, and Age Patterns

Significant physical and demographic shifts are under way and will continue in the United States. Migration for the preceding decade has been from the Northern industrial states to the sunbelt states, the South, and the Southwest. Moreover, the population trends of the 1950s and 1960s indicate that in the 1980s more entry-level jobs will be available than workers to fill them. Additional considerations will have to be given to the opportunity for effectively utilizing highly skilled elderly people who, because of longer life expectancies, will be available and willing to work. Working at home is another viable alternative to consider for certain types of corporate work.

Productivity Expectations

As business competition increases and as the U.S. economy is restructured to accommodate shifting populations, income redistribution, and the growth of service sector industries, there will be a growing awareness of and concentration on increased productivity opportunities. This awareness, coupled with the ongoing expansion, availability, and lower cost of technology will continue to drive the expansion of office and other forms of automation.

Technological Innovation

The decade of the 1980s will witness an ever continuing development and introduction of new technology and related implementation opportunities. As production costs of the new technology continue to decrease because of improved manufacturing and development techniques, application opportunities considered unfeasible the preceding year will become cost justified in the current year. This trend will continue throughout the decade, and miniaturization and portability will have a significant impact on every aspect of the work place.

Inflation

Direct costs associated with labor will continue to increase at inflationary rates because in many cases salaries are closely linked to the consumer price index (CPI). In addition, we will continue to witness a growing militancy among white-collar workers for salary and benefit demands related to and above the CPI. This trend is clearly evidenced by the rapid increase in white-collar union membership over the past decade (see Figure 10-2).

As the above-mentioned factors drive office automation and as the technology improves productivity within the office, the way in which secretarial, clerical, administrative, managerial, and professional employees work and accept the new work roles that will develop will also be affected. Only through awareness and sensitivity to the resultant societal, organizational, educational, and environmental issues can the office systems planner avoid serious obstacles.

Issues related to office systems, such as increased job specialization, displacement and/or reeducation of middle-management

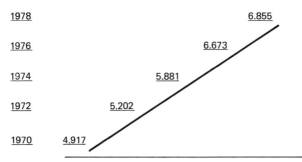

Figure 10-2 Number of white collar union members (million). [U.S. Bureau of Labor Statistics (1978).]

staff, organizational stratification, increased span of control, and depersonalization, will have to be carefully evaluated, understood, and resolved. If these issues are not properly addressed, office systems will be poorly accepted and most likely rejected.

On the basis of experience with computer systems, word processors, and other office technologies, it has been generally observed that people tend to be complacent and generally resist change. There are many reasons for this resistance, but as John Connell has pointed out, "They do so because they perceive machine-based systems to be rigid, structured, unforgiving of errors and unresponsive to changing business conditions. Systems have generally been designed to make processing machinery run more efficiently rather than respond to the needs of users."[2]

Obviously then, a good deal will have to be done to overcome staff resistance to technology. Solutions will require joint management and user study analysis, which will allow more open selection, development, and implementation of office systems technology. As Connell has stated regarding the implications of the new office technologies:

The real challenge in the office of the 80's is not technology, but behavioral. Although there are some gaps in the technology, one can be confident that they will be filled in due course. Based on the record of systems efforts to date, less assurance can be given that behavioral concerns will be addressed unless senior management makes them an integral part of a program.[3]

It is important to recognize that productivity measures and efforts must never result in an unsatisfactory working environment if they are to be successful. Office systems will have to be introduced in a manner that will not intimidate or dehumanize. Anxiety and concern with office system implementation occur when no consideration is given to established office operating procedures or to the relationship between people and groups.

As office systems automation becomes more prevalent and pervasive, it will become apparent that a number of clerical and administrative office tasks will have to be redefined, reclassified, reevaluated, and established into the context of "new" administrative functions with associated higher pay, job description, career paths, and recognition as a valued member of the corporate community.

As James Driscoll has pointed out, "Office employees are motivated to stay with the organization and perform effectively when they feel there is a good fit between them and their individual needs and the opportunities provided by the organization."[4] Experience indicates that individual secretaries and clerks differ substantially in personal preferences. Some seek advancement to higher organizational levels; others want an opportunity to cultivate and use particular technical skills; still others need only a secure job and a source of income among congenial co-workers. Obviously it will be necessary to be sensitive to and aware of the issues of career path development. Similar thinking will also have to be applied when dealing with managerial and professional workers.

Formal systems for identifying the needs of individual employees and matching them with organizational opportunities must be developed. Establishing counseling sessions, posting job opportunities, and allowing open job bidding reflect this orientation. Clear career paths must be made obvious, in order to satisfy personal goals and the needs of the different individuals who work in the office.

A major responsibility of the office systems planner will be to point out the various information processing and other emerging trends and to translate these trends into promising opportunities. It will be a formidable challenge to face the task of integrating the diverse and often complex technologies into a cohesive plan for office automation.

ENVIRONMENTAL AND ERGONOMIC ISSUES AND CONCERNS

The introduction of office system technologies into the general work area will change considerably the physical planning of office systems. Elements of the office that were not previously considered when the data processing function was totally centralized will now be reviewed with considerable care and diligence, and the office systems planner will be faced with both new opportunities and serious concerns. The considerations will include such areas as air conditioning, lighting, security, space planning, electrical distribution facilities, and acoustical requirements.

The design of office systems and its adaptability to the general working environment have recently been the subject of much attention and discussion. Although the issues are complex, the general human/machine interface issues may be broadly categorized as follows:

Screen glare, luminance, color, flicker

Character generation/size/spacing

Furniture

Temperature and humidity levels

Radiation emissions

Noise

Lighting

Static electricity

Terminal exteriors and positioning

Maintenance/cleaning

Paperwork and source document legibility

In line with the office environment, the office systems planner must have a solid knowledge and grasp of the physical, environmental, and ergonomic issues associated with future systems implementation and planning. These considerations include the following:

Space quality. Square areas are most desirable. Long, narrow areas present traffic problems, increase reflective noise levels, and significantly reduce system flexibility. Irregular walls and surfaces should be avoided wherever possible. High ceilings are particularly desirable, because the noise level is significantly reduced.

Space quantity. It is particularly important that careful consideration be given to anticipated growth. Factors to be considered should include staff and organizational needs, and plans and projections of future system requirements.

Structure. The planner should be aware of and sensitive to floor loading and capacity limitations. Close coordination and communication with facilities management will avoid serious floor-loading problems. Whenever wall mounted systems components are used, it is also important to be aware of the types of fasteners utilized, so that safety considerations are properly addressed.

Power. Power systems may be distributed in several ways. The three most common forms are flexible distribution grids in ceiling plenums, raised flooring grid systems, and the more usual floor distribution systems. In using any of these systems it is important to understand the initial versus the long-term cost options and complications. The office systems planner must have a complete understanding and knowledge of system requirements in evaluating power distribution requirements.

Telephone/communications. The planner must be aware of the type of installed telephone systems so that duct utilization and cable distribution may be properly evaluated. Where appropriate, the telecommunications manager or telephone company marketing-/engineering representative should be consulted.

Acoustical considerations. Control of sound within the operating system environment is the primary objective of an acoustical review. Several methods are available, and they may be used in combination. The most common technique involves sound masking. Whenever possible, highly sound-absorbent surfaces should be used to reduce sound reflection. Ceilings, carpets, and vertical structures are all of prime importance in reducing sound. The planner should be aware of special ceiling products specifically designed to reduce sound.

Lighting. Conventional lighting is usually integrated with the ceiling grid system. This approach offers limited flexibility and is somewhat inefficient. Indirect lighting places light where it is required, and is superior in terms of the quality and quantity of light provided. It is particularly beneficial in meeting temporary system requirements.

Air conditioning (HVAC). Both human engineering and system requirements dictate that careful attention be given to heating and cooling of the office environment. Research has indicated that no single design will meet the requirements of everyone. To overcome the heat given off by systems, people, lighting, and surrounding surfaces, the air supply must be carefully regulated and controlled through cooling, or the heat may be removed through a requirement that air be circulated in a room or a floor a certain number of times per hour.

Many of the issues discussed above have been satisfactorily resolved in the factory over time. When encountered in the office, they should be considered a challenge to be overcome, not an impediment to progress.

Results to date have not revealed any major or significant health hazards as a result of office systems installations. Some studies have revealed, however, that visual and postural discomfort may result when equipment design or related factors are deficient in the working environment. Consequently, office systems designers must become thoroughly familiar with such issues and must be prepared to take appropriate actions and measures to avoid environmental problems. The installation of any current or future technologies will have to be subject to careful review of health, safety, environmental, and ergonomic issues and must be the constant concern of those responsible for future system implementations.

FUTURE TECHNOLOGICAL IMPLICATIONS

When the technological changes of the past several years are reviewed, it seems that decades of development have been merged into years and sometimes into months.

Thinking about this rapid change enables the office systems planner to visualize the directions and related implications that new technology will have over the next several decades.

As we enter this new information age, it is apparent that technology will have a significant impact on the entire structure of our society—the way we work, how we engage in travel and sports, how we educate, think, and even sleep and eat.

The period of change over the next decade will be rapid, dramatic, and exciting. To gain a perspective on the size of this rapidly growing marketplace, consider the following statistics, compiled by the consulting firm of A. D. Little Inc. Little projects that computer growth will continue in the United States at 8% to 9% a year. The firm projected a U.S.-installed computer base, by 1982, with a value of $76 to $80 billion dollars. In terms of office information systems, Little projects that by 1985, $15 billion will be expended annually.[5]

Concurrent with this growth, the percentage of the labor force that depends on data processing for support will increase from the present 30% to 40% to 70% by 1985. As these changes occur and as data processing and office system technologies continue to become major segments of the economy and our society, the planner must constantly question how the technologies will be beneficial. It will be necessary to plan effectively and coherently to introduce change in an acceptable way suitable for the existing business environment.

Technological Proliferation

Let us consider the specifics of the technological proliferation and change we are currently witnessing and are likely to witness over the next decade.

Advancing technology and lower manufacturing costs will provide enormous new opportunities for the use of technology in ways never before considered. The driving force will be continued improvements in the development of very large scale integration (VLSI) technology. The improvements will allow manufacturers to place many thousands and even millions of transistor circuits on a single chip of silicon or some similar material. These capabilities will result in exponential improvements in small system performance and rapid deployment of the new technology capabilities into the everyday work place. Similar improvements will occur in the development of magnetic storage devices as capacity increases and costs diminish. In the 1960s, it cost about $25 to $30 to store a million characters of information. Today, the same information can be stored for less than 40 cents.

As the very large scale improvements and the lower storage costs develop, we will witness rapid movement toward real-time, on-line processing. This will be particularly evident in the office environment. Such rapid change is already evident in the transactional environment as these new technologies are introduced into supermarkets (universal product code scanners) and into banking (automated teller systems).

We have recently witnessed rapid cost reductions in all three categories of computer systems: mainframes, minicomputers, and microprocessors.

As the computer industry moves toward systems that are smaller, faster, and cheaper, we can expect a rapid proliferation of microprocessor products capable of performing a remarkably broad range of specialized tasks at extremely low cost. A review of the costs associated with personal calculators and digital watches reveals the trends we should expect in small business and personal computer system costs.

Computer peripheral equipment will also continue to decrease in cost and increase in functionality. We can expect a dramatic increase in the number and types of available terminals. The major attraction of many of these new products will be the availability of local intelligence that will provide users with unique application software to meet individual and local requirements, and provide access to remote computer facilities for additional information or processing requirements. Another growth area will be the use of portable terminal units capable of remote communication with distributed or host computers, or external information utilities.

Of particular importance will be the communications evolution we will witness over the decade. Already we are offered broadband communications capability that provides opportunities to transmit voice, data, text, and image information simultaneously. As high-speed satellite communications facilities increase, current high costs will decrease as volume use increases. By the mid-1980s, use of these services by business will become fairly commonplace. In the latter part of the decade, these services will be available at modest cost to all businesses and to many homes. We are now witnessing this in industries such as radio, television, newspaper, magazine, book publishing, movie, and entertainment, which provide information about new products and services on an almost daily basis.

We can expect to continue to see rapid advances in printing technology. Ink jet, xerographic, and laser techniques will provide opportunities to produce imaginative and creative computer output. Electronic photography will provide options to store, recall, transmit, edit, and review images in color. These images will then be positioned to be merged with text and transmitted to printers, digital typesetters, or other devices for printing or platemaking. Major advances consolidating word processing, data processing, communications, and typesetting capabilities into individual systems will be realized in the early 1980s. These systems will provide electronic editing, retouching, color correcting, sizing, and cropping of information and photographs on a single system.

In terms of software products, we should continue to expect a migration away from traditional languages such as FORTRAN and COBOL, toward user-oriented, easy-to-use languages such as BASIC, PASCAL, ADA, and others. In addition to ease-of-use factors, software cost considerations, whether due to in-house development efforts or external sources, will be in the billions of dollars. Recent estimates show that in the United States more than $20 billion per year is spent on programming staffs alone. Besides easier-to-use software languages, more advanced operating systems, data-base management offerings and structured techniques will reduce software development time and expense. The office systems planner should anticipate that much software will be "hard-wired" directly onto individual chips, with microprocessors providing operating system support. The planner will be constantly challenged to plan effectively and coherently to introduce the resultant products in an acceptable manner appropriately suited for the socioeconomic conditions of the business.

Technological Planning Issues

The office systems department will have to evaluate the technological trends discussed above in terms of integrated systems planning. This means that:

1. Systems planning will require greater integration of systems skills, personnel abilities, and related resources. More integration of different types with multilevel resources and a better understanding of business solutions will be required.

2. In the past five years, many multidivisional businesses have adopted strategic planning concepts with specific objectives and management control systems for tracking performance against objectives. Obviously, a good information management system will have to capture and report information either internally generated or available on an external data base. This is required by the business to monitor its performance against the strategic plan. Consequently, to a much greater extent in the future, the process of planning of information systems will be synchronized with the business strategic planning process.

3. Information system users will have to become much more involved in information system planning. Therefore, the planning process will have to be more user oriented. Managers in user departments are becoming more comfortable with computers than they were 10 years ago and must therefore participate to

a greater extent in the planning process. Further, as distributed processing moves more into user departments and as users have more direct contact with computerized systems, we can expect that users will be more demanding in terms of what applications are computerized and how the systems will work.

Obviously, a new set of learning experiences will emerge as we increase our knowledge about the flow of information and data manipulation requirements, and then match these needs and requirements to available technologies. As products are planned, tested, and implemented, the serious issue of proper organizational structure and responsibilities will have to be reassessed accordingly. At the same time, the office systems planner must demonstrate the personnal qualities necessary to accomplish the job at hand and establish a positive profile in the eyes of management.

The economic, societal, and business implications of these office products and services are awesome and should be a constant concern to the office systems planning department, because they will impact every area of business activity.

Besides all the negatives and potential for problems related to the technological evolution, with the proper planning, introduction, education, and training, these difficulties can be significantly reduced. As people start to feel more comfortable with systems and recognize the potential for enhancing their capabilities in business or enriching their lives at home, the new technologies will be more readily accepted.

One futurist has stated, "What we are creating is a new class structure around wealth—this time the wealth of information. Like today's 'haves and have-nots,' we will be a society of 'knows and know-nots.' "

As we reach conclusions about the future, certain things are obvious. The office and the home as we know them today will change radically as a result of communications-based technology. Less reliance will be placed on the "work place" concept of today. Information distribution will permit work to be brought to wherever the worker happens to be.

The 1980s will be a decade of rapid change and excitement to those deeply involved in developing, shaping, and using the emerging office technologies. There will be great pressures, some defeats, and, it is hoped, many victories. It will be a turbulent decade presenting many great opportunities.

KEY IDEAS

Over the next decade of the nineties, administrative problems and technical solutions will become much more complex, and participative management will create behavioral, societal, and technological issues that will have to be addressed and resolved by those responsible for office automation.

In addition to these responsibilities, planners must have a broad and thorough understanding of such issues as:

Energy shortages

Educational levels

Government and regulatory reporting requirements

Shortages of skilled personnel

Demographic work, life, and age patterns

Productivity expectations

Technological innovations

Inflation

These issues must be considered through well-thought-out strategies and plans that anticipate change. A major consideration will involve behavioral patterns that will occur in the office environment as technology proliferates.

Particular attention will have to be addressed to environmental and ergonomic issues, such as

Space quality and quantity

Structure

Power

Communications

Acoustics

Lighting

Air conditioning

From a technical perspective, it will be necessary to have a complete understanding of the various available and emerging technologies, as well as a masterful understanding of business and functional needs and requirements.

The following decades will involve significant organizational, behavioral, and technical changes. This period will also offer outstanding opportunities for satisfaction, progress, and legitimate enjoyment to those associated with office automation and the emerging information resource field.

REFERENCES

1. Howard L. Morgan, "The Future of the Office of the Future," Presentation Paper, 1980 NCC Office Automation Conference, Houston, Texas.

2. John J. Connell, "Office of the 80's—Productivity Impact," Office Technology Research Group, Special Advertising Section, 1980.

3. Connell, *ibid.*

4. James W. Driscoll, "People and the Automated Office," *Datamation,* December 1979.

5. N. S. Zubel, R. L. Frank, and A. M. Mayfield, "The Emerging Real World of Office Automation," Report, Arthur D. Little, Boston, MA, June 1979.

APPENDIXES

APPENDIX A

Position Descriptions

Table A-1   Systems Specialist

POSITION TITLE: SYSTEMS SPECIALIST

DIVISION/DEPARTMENT: MANAGEMENT INFORMATION
SYSTEMS

MANAGER'S TITLE: DIRECTOR—OFFICE SYSTEMS

NEXT LEVEL OF MANAGEMENT: VICE PRESIDENT—MIS

PROFESSIONAL REPORTS: __   NONPROFESSIONAL REPORTS: __

BUDGET: $XXX,XXX   LOCATION: CITY, STATE

PROJECTS

Develop standards for office systems related equipment.

Establish an inventory control system for office automation.

Develop a five-year plan for word processing, filing, photocompostion, OCR, and other related systems.

Perform the equipment evaluation function for users ordering new systems.

Establish selection criteria and methodologies for vendor evaluation.

PRINCIPAL RESPONSIBILITIES

Evaluate current technology with regard to office systems and develop a strategic plan based on the integration of word processing, filing, photocomposition, data processing, OCR, and other related technologies.

Assist users in the evaluation of vendors of word processing and related systems and establish criteria to facilitate selection.

Establish and maintain a corporate inventory of office systems, beginning with word processing equipment.

Affiliate with appropriate user groups in order to become current in office automation technology.

SIGNATURE OF APPLICANT: _____   DATE: _____

MANAGER'S APPROVAL: _____   DATE: _____

NEXT LEVEL OF MANAGEMENT APPROVAL: ___   DATE: _____

220

Table A-2   Communications Specialist

POSITION TITLE: <u>COMMUNICATIONS SPECIALIST</u>
DIVISION/DEPARTMENT: <u>MANAGEMENT INFORMATION</u>
<u>SYSTEMS</u>
MANAGER'S TITLE: <u>DIRECTOR—OFFICE SYSTEMS</u>
NEXT LEVEL OF MANAGEMENT: <u>VICE PRESIDENT—MIS</u>

PROFESSIONAL REPORTS: __   NONPROFESSIONAL REPORTS __
BUDGET: <u>$ XXX,XXX</u>   LOCATION: <u>CITY, STATE</u>

PROJECTS

Develop a five-year communications plan.

Determine the communications requirements for integrated office systems.

Design a prototype in-house electronic mail system.

Evaluate the communications aspects of user word processing and electronic mail systems.

Isolate a standard for intracompany and intercompany communications.

Become familiar with communications standards for office systems and regulatory issues regarding electronic mail.

PRINCIPAL RESPONSIBILITIES

To cooperate with the communications department in an effort to integrate the communications requirements for future office systems within the overall corporate communications plan—develop a five-year office systems communications plan.

To evaluate user requests for communications support utilizing either communicating word processing or electronic mail technology and to design a prototype in-house electronic mail system.

To isolate a standard communication protocol for intracompany and intercompany communications and to become current in the efforts of standards committees and regulatory issues concerning electronic mail.

SIGNATURE OF APPLICANT _____ DATE _____
MANAGER'S APPROVAL _____ DATE _____
NEXT LEVEL OF MANAGEMENT APPROVAL _____ DATE _____

Table A-3   User Specialist

POSITION TITLE:	USER SPECIALIST
DIVISION/DEPARTMENT:	MANAGEMENT INFORMATION SYSTEMS
MANAGER'S TITLE:	DIRECTOR—OFFICE SYSTEMS
NEXT LEVEL OF MANAGEMENT:	VICE PRESIDENT—MIS

PROFESSIONAL REPORTS: __   NONPROFESSIONAL REPORTS __

BUDGET: $XXX,XXX   LOCATION: CITY, STATE

PROJECTS

Develop user study methodology relating to secretarial, clerical, professional, and managerial activities and interrelationships. Develop study methodology necessary to determine corporate mail practices and requirements.

Develop study methodolgy necessary to determine corporate filing requirements.

Assist users in determining word processing and electronic mail requirements.

PRINCIPAL RESPONSIBILITIES

Develop the survey methodology and evaluation techniques necessary to understand what different employees do and how much time they spend doing it. Include interviews, data collection, random observations, analysis, and observation techniques.

Test these methodologies and compare results with the published results of researchers.

Participate in user studies to determine the cost justification of word processing and other office systems.

Develop potential cost justification models for senior management review.

SIGNATURE OF APPLICANT _____ DATE _____
MANAGER'S APPROVAL _____ DATE _____
NEXT LEVEL OF MANAGEMENT APPROVAL _____ DATE _____

Table A-4   Human Resources Specialist

POSITION TITLE: <u>HUMAN RESOURCES SPECIALIST</u>

DIVISION/DEPARTMENT: <u>MANAGEMENT INFORMATION</u>

<u>SYSTEMS</u>

SUPERIOR'S TITLE: <u>DIRECTOR—OFFICE SYSTEMS</u>

NEXT LEVEL OF MANAGEMENT: <u>VICE PRESIDENT—MIS</u>

PROFESSIONAL REPORTS: __   NONPROFESSIONAL REPORTS: __

BUDGET: $_____   LOCATION: <u>CITY, STATE</u>

PROJECTS

Develop methodology necessary to study effects of various technologies on personnel.

Study different organizational implementations of word processing and electronic mail.

Participate in the study of interoffice relationships.

Assist in the design of future office systems.

Assist in facility planning activities

PRINCIPAL RESPONSIBILITIES

Develop the methodologies required to evaluate the effects that office systems may have on users and their organizations. Consider job satisfaction, morale, turnover, and productivity.

Assist in the design of future systems from the ergonomic perspective and the man/machine interface aspect.

Affiliate with research organizations that specialize in office environment considerations, especially health and safety.

SIGNATURE OF APPLICANT: _____   DATE: _____
MANAGER'S APPROVAL: _____   DATE: _____
NEXT LEVEL OF MANAGEMENT APPROVAL: ____   DATE: _____

Table A-5   Facilities Specialist

POSITION TITLE: FACILITIES SPECIALIST	
DIVISION/DEPARTMENT: MANAGEMENT INFORMATION SYSTEMS	
SUPERIOR'S TITLE: DIRECTOR—OFFICE SYSTEMS	
NEXT LEVEL OF MANAGEMENT: VICE PRESIDENT—MIS	

PROFESSIONAL REPORTS: __	NONPROFESSIONAL REPORTS: __
BUDGET: $_____	LOCATION: CITY, STATE

PROJECTS

Develop a five-year plan pertaining to the impact that office systems will have on the environment.

Design and implement site preparation checklist for each office systems site and building/campus site(s).

Coordinate major word processing, and other electronic mail equipment and other installations between hardware, communications, electrical equipment and facilities.

Develop environment standards for office systems.

PRINCIPAL RESPONSIBILITIES

Contribute to the five-year plan from the perspective of electronic office systems becoming more and more prevalent in the office. Consider the potential for the integration of environmental controls within office systems.

Design a standard site preparation package for office systems with a supplement for specific systems. Include electrical and communications requirements along with ordering and delivery estimates.

Assist users with the facility planning for large word processing and electronic mail systems.

Develop standards for office system components to ensure high quality and adherence to fire and safety standards.

SIGNATURE OF APPLICANT: _____	DATE: _____
MANAGER'S APPROVAL: _____	DATE: _____
NEXT LEVEL OF MANAGEMENT APPROVAL: ___	DATE: _____

APPENDIX B
Functional Matrices

Table B-1 Automated Office Systems: Summary of Functional Matrices

Job Type	Overall Potential	Creation		Processing (Change)		Distribution (Dissemination)		Retention (Search Retrieval, Disposal)	
		Potential	Function	Potential	Function	Potential	Function	Potential	Function
Chief executive officer (CEO)	Low	Low	Speeches Presentations Decisions	Low	—	Low	—	Low	—
Vice president (VP)	Modest	Low	Reports Correspondence Decisions	Modest	—	Modest	—	Low	—
Upper management (general manager/ director)	Moderate	Moderate	Reports Schedules Correspondence Decisions	Moderate	Productivity data Cost data Budgets Forecasts/ plans What-if analysis	Moderate	Reports Strategy Plans Orders	Good	Data Various documents

Middle management (director and manager)	Good	Moderate	Reports Orders Decisions	Good	Productivity data Budgets Costs Status reports Correspondence What-if's	Good	Reports Orders	Good	Data Various documents
Professional /paraprofessional/ technical	High	Good	Reports	Good	Various data	High	Reports	High	Documents
Executive secretary (administrative and nonadministrative)	Moderate	Good	Schedules Reports Correspondence Telephones	Moderate	Schedules Correspondence Mail Telephone	Good	Various	Moderate	Documents Reports
Manager secretary (general manager/director)	High	Good	Schedules Reports Correspondence Telephones Dictation	High	Schedules Budgets Correspondence Reports Mail Telephone	Good	Various	High	Reports Data Various data

Table B-1 Automated (Continued)

Job Type	Overall Potential	Creation		Processing (Change)		Distribution (Dissemination)		Retention (Search Retrieval, Disposal)	
		Potential	Function	Potential	Function	Potential	Function	Potential	Function
Administrative secretary (director/manager)	High	Good	Dictation Reports Verbal	High	Reports Schedules Mail Telephone Orders	High	Various	High	Documents Reports
Office worker	High	Good	Reports Correspondence Verbal	Good	Data Reports	Good	Various	High	Files
Clerk/typist	Good	Good	Dictation Handwritten reports	Good	Various	Good	Various	Low	Files

Table B-2 Office Systems: Functional Matrix—Information Creation (Includes Capture & Preparation)

Job Type	What	From	How	Frequency	Volume	Potential
Chief executive officer (CEO)	Policy statements Public statements Directives/orders/regulations Personal correspondence Operating statements	Stockholders Board members Financial and commercial Government agencies Associates	Verbal Mail* Telephone* Reports* Conferences*	Multiple times/weeks	Low	Low
Administrative executive secretary	Schedules/calendar Letters/memos Meetings	CEO Subordinate(s)	Handwritten Typed* Telephone*	Active and on demand/reactive	Modest	Good
Vice president (VP)	Policy Procedures Directives/orders Personal correspondence	CEO Organization reporting to him/her Peers	Verbal Review/approve Mail* Reports* Conferences*	Multiple times/day	Modest	Modest
Executive secretary	Schedules/calendar Letters/memos Reports Follow-up	Vice president	Typed* Telephone* Verbal Dictation*	On demand/reactive	Moderate	Good

Table B-2 *(Continued)*

Job Type	What	From	How	Frequency	Volume	Potential
Upper management (general manager/ director)	Reports Orders Memos Personal correspondence Strategies	Vice president/ director Peers Reporting Organization	Verbal Reports* Mail* Telephone* conference*	Multiple times/day	Moderate	Moderate
Director/ general manager secretary	Schedules Letters/memos Minutes Reports	General manager/ director	Typed* Telephone* Verbal* Dictation*	On demand/ reactive	High	Good
Middle management (director/ manager)	Reports Operational decisions Orders Memos	Superiors Peers Reporting Organizations	Handwritten Verbal Reports* Mail* Conferences*	Multiple times/day	Moderate	Good
Adminis- trative secretary	Schedules Reports Documents for Retention	Professional Paraprofessional	Typed* Telephone* Handwritten Dictation*	On demand/ reactive	Moderate	Good

Professional/ Paraprofessional/ technical	Reports Proposals	Superiors Peers	Reports* Mail* Verbal	On demand/ reactive	Moderate	High
Office manager	Reports	Superiors	Typed* Handwritten* Verbal	On demand	High	High
Clerk/typist	Various (few)	Various	Typed* Verbal	High	High	Modest

*Functions being addressed by the automated office of the future.

Table B-3 Office Systems: Functional Matrix Process—Processing (Includes Change and Expansion)

Job Type	What	How	Frequency	Volume	Potential
Chief executive office (CEO)	Policy items Strategies Financials	Assimilation Creativity Mental	Multiple times/month	Low	Low
Administrative executive secretary	Expedites/routes Reports Coordinate responses Schedules Priorities	By hand* Typed* Copy*	Active/reactive on demand	Modest	Moderate
Vice president (VP)	Budgets/books Strategies/plans	Mental What if-ing	Multiple times/month	Low	Moderate
Executive secretary	Expedites/report Schedules Mail priority Phone priority	By hand* Typed* Copy*	Multiple times/day	Moderate	Good
Upper Management (General manager/director)	Productivity figures Budget/costing data Forecasts/plans Mental/what-if-ing	CPU application* Calculator* By hand* Mental	Multiple times/month	Modest	Moderate

Role	Tasks	Methods	Frequency		
General manager/director secretary	Expedites reports Schedules Mail priority Telephone priority	By hand* Typed* Copy*	Multiple times/day	Moderate	High
Middle management (Director/manager)	Productivity figures Cost data Industry reports Status reports	CPU application* Calculator* By hand* Mental	Multiple times/month	Moderate	Good
Administrative secretary	Expedites reports Schedules Mail phone priority Orders	By hand* Typed* Copy*	Multiple times/day	High	High
Professional/paraprofessional/technical	Raw data from various sources	By hand Calculator* Terminal* Personal composition* Mental Copy*	Multiple times/day	High	Good

Table B-3 *(Continued)*

Job Type	What	How	Frequency	Volume	Potential
Office worker	Various numbers/data	By hand* Calculator* Copy*	Multiple times/day	High	High
Clerk/typist	Little	Typewriter*	High	High	High

*Functions being addressed by automated office of the future.

Table B-4 Office Systems: Functional Matrix—Distribution (Includes Dissemination)

Job Type	What	To	How	Frequency	Volume	Potential
Chief executive officer (CEO)	Annual reports Policies Personal correspondence Speeches	Stockholders Board members Financial/industry Community Government agencies Accountants	Printed report* Personal stationery* Company letterhead* Personal notes* Verbal Dictation*	Multiple times/month	Low	Low
Administrative executive secretary	Various	Various	Verbal Type* Telex*	Multiple times/day	High	Moderate
Vice presidents (VP)	Policies/guidelines Orders Speeches Personal correspondence Associates	CEO Peers Reporting organizations Industry/community Personal dictation	Personal stationery* Company letterhead* Personal notes* Verbal	Multiple times/month	Low	Low
Executive secretary	Various	Various	Verbal Typed*	Multiple times/day	High	Good

235

Table B-4 (Continued)

Job Type	What	To	How	Frequency	Volume	Potential
Upper management (general manager director)	Reports Orders Strategies Memos Personal correspondence	Superiors Subordinates Peers/associates	Company letterhead* Memo paper* Verbal Personal dictation*	Multiple times/week	Modest	Moderate
Manager secretary (General manager/ director)	Various	Various	Verbal Typed*	Multiple times/day	High	Good
Middle management (director/ manager)	Reports/analysis Orders	Superiors Subordinates Peers/associates	Computer printout* Verbal* Dictation*	Multiple times/week	Moderate	Good
Administrative secretary	Various	Various	Verbal* Typed* Handwritten Telex* Facsimile*	Many times/day	High	High

Profes-sional/Paraprofes-sional, Technical	Documents/manuals Reports	Superiors Subordinates	Computer output* Verbal	Multiple times/week	High	High
Office worker	Various	Superiors	Handwritten Typed*	Many times/day	High	Good
Clerk/typist	Letters Memos	Author	Typed*	Many times/day	High	Good

*Functions being addressed by automated office of the future.

Table B-5   Office Systems: Functional Matrix—Process Retention (Includes Search, Storage, Retrieval and Disposal)

Job Type	Source	What	How	Amount	Maximum Speed of Referral Needed	Needed Duration	Actual Duration	Potential
Chief executive officer (CEO)	Board of directors SEC/ Government Agencies Industry/ professional & management organizations	Reports/ articles Personal correspon- dence	File cabinets Security— good	Low	Fast	1–12 months	2–4 years	Low
Adminis- trative executive secretary	Superior	Correspon- dence (active) Reports Records Schedules	File cabinets* Security— good	Low	Fast	1 day–12 months	1–4 years	Modest

Vice president (VP)	Reporting organization Industry associations Peers	Performance reports Strategy documents Personal correspondence Financial reports	File cabinets Security—good	Low	Within 15–60 minutes	1–12 months	2–4 years	Modest
Executive secretary	Superior	Correspondence Reports Records Schedules	File cabinets* Security—good	Low	Fast	1 day–12 months	1 day–4 years	Moderate
Upper management (general manager/ director)	Superiors Subordinates Peers Industry	Budget/cost data Personnel records Performance data Project reports Industry reports Correspondence Miscellaneous	File cabinets* Security—modest	High	15 minutes	6–24 months	1–3 years	Moderate

Table B-5 *(Continued)*

Job Type	Source	What	How	Amount	Maximum Speed of Referral Needed	Needed Duration	Actual Duration	Potential
Manager secretary Director/general manager	Superior	Correspondence Reports	File cabinets* Security—good	Low	Within 5–45 minutes	1 day–12 months	1 day–4 years	Good
Middle management (director/manager)	Superiors Subordinates Peers Industry	Budget/cost data Personnel records Performance data Project reports Policy/technical manuals Correspondence	File cabinets* Security—poor On-line storage Archives* Vaults	High	Within 15 minutes	3–24 months	1–2 years Up to 6 years for vital records	Good

Professional/ paraprofessional/ technical	Superiors Peers Industry	Raw Data Reports/ correspondence Project data	File cabinets* On-line storage* Security— poor	Very high	Within 1 minute Fast as needed	1–6 months	1–2 years	High
Administrative secretary	Various	Various	File cabinets* Security— poor	Very high	Fast	1 day–12 months	6–24 months	High
Office worker	Various	Various	File cabinets* Security— poor	Moderate	Moderate 5–60 minutes	1 day–6 months	6–24 months	High
Clerk/typist	Various	Various	File cabinets* Security— poor	Low	Within 6 hours	1 day–2 months	1 week–1 year	Good

*Functions being addressed by automated office of the future.

241

Management, Principal, Secretarial, & Clerical Questionnaires

Table C-1 Managment/Exempt Questionnaire

I. Background Information

    1. Name: _____

    2. Title: _____

II. Job Responsibilities

Please describe your major responsibilities. List up to five, in order of their importance, or attach a copy of your latest job description.

    1. _____

       _____

       _____

    2. _____

       _____

       _____

    3. _____

       _____

       _____

    4. _____

       _____

       _____

    5. _____

       _____

       _____

244

Table C-1 (*continued*)

III. Time Distribution by Activity and Travel

1. Please indicate the average amount of time (in hours or minutes) you spend *each week* in each of the following activities (include time spent performing these activities at home, while commuting, etc., but *exclude* activities performed while traveling on business):

   _____ telephone—scheduling meetings

   _____ telephone—other

   _____ calculating

   _____ copying

   _____ meetings

   _____ operating a computer terminal

   _____ typing (exclude computer terminal typing)

   _____ writing

   _____ dictating to a secretary

   _____ dictating to a machine

   _____ proofreading

   _____ reading

   _____ incoming-mail handling

   _____ file handling (locating, retrieving, updating, purging, refiling)

   _____ other paper handling (sorting, collating, binding, discarding, etc.)

   _____ searching for information (by telephone, in files, etc.)

   _____ reviewing programs/reports related to computerized systems

   _____ other major activities: _____

   _____

   _____

Total   _____

2. Please indicate the average amount of time (in hours or minutes) you spend *each week* in each of the following general activities:

   _____ Personnel matters that relate to you or your staff (performance reviews, interviews, evaluations, counseling, etc).

   _____ Administrative activities coordinated/requested by the administration/education/planning function

Table C-1 (*continued*)

_____ Profit Plan

_____ Standards preparation, review, revision

_____ Project Reporting (status, cancellations, changes, estimates, etc.)

_____ Other major activities

_____ Project management/reporting activities requested/conducted by you, your management, and/or user personnel to benefit your group or a user group.

3. How frequently do you travel in a year? _____

On the average, how long are your trips? _____

IV. **Labor Intensive and Peak Workload Activities**

1. Please list below *major* labor intensive manual activities that you perform. (Note: Project leaders should include activities performed by staff members as well, i.e., record keeping, data preparation, report preparation, calculating, etc.)

_____

_____

_____

_____

_____

2. Do you have peak work-load periods? Yes_____No_____
   If yes:

   When?   How Long?   What are the reasons?

_____

Table C-1 (*continued*)

_____

_____

_____

_____

V.   Delegable Tasks

Please list any tasks you now do that could be delegated if more support were available.

Tasks that could be delegated	Estimated number of hours you could save by delegating these tasks		To whom could these tasks be delegated?
	Hours/Day	Hours/Month	Secretary Administrative Specialist

List the most important activities that you would accomplish with the time made available from delegating the tasks listed above, and your best estimate of the value of these.

List of Activities	Estimated Value of Activities (Expense Reductions, Staff Reductions, etc).

Table C-1 (*continued*)

VI.   Documents Received—Computer Generated

List the most important documents that you receive.	Source received from	Average number of packages	Briefly describe how you use each of these documents and why each is of value	Document retention period	Is information in this document used to create another document? (Yes/No)	Briefly describe any problems in using/ receiving information in those documents

---

VII.   Additional/Miscellaneous Comments

Table C-1 (*continued*)

Table C-2 Principal* Questionnaire

Note: A principal is anyone who needs, has, or requests secretarial, administrative, or clerical support).

1. Name _____

2. Title _____

3. Department _____

4. Location _____

5. Work phone number and Extension _____

6. Name(s) of your secretary:

    _____

    _____

    _____

    _____

7. Number of years in the profession: _____

8. Number of years in this company: _____

9. Number of years in this position: _____

10. Fill in the blanks with the appropriate letters in regard to the following equipment:

    a. Located in my own office and I use it.
    b. Located in my office, but I do not use it.
    c. Have access to and I use it.
    d. Have access to but do not use it.
    e. Equipment not available.

    _____ telephone
    _____ mechanical telephone-answering device
    _____ speaker phone

Table C-2 (*continued*)

_____ intercom

_____ calculator or adding machine

_____ computer terminal

_____ Telex or TWX®

_____ individual dictation unit

_____ dictation system (one with a remote recorder)

_____ facsimile or telecopier

_____ copier

_____ typewriter

11. Can you type?_____ (Yes/No)

12. If you do any typing at work, check the kind(s) of typewriter you use.

_____ video display or CRT typewriter

_____ magnetic card, tape, or memory

_____ manual, electric, or correcting

_____ other (specify)

_____ none

13. If you do not have a typewriter, would you like one?_____ (Yes/No)

14. Check the following business services you use from other companies:

_____ messenger service

_____ graphics/composition

_____ printing

_____ consultants/professionals

_____ computer time

_____ typing service

_____ office temporaries

_____ other (specify) _____

15. If you use messenger services, what percent of the time of each do you use? (The total should equal 100%.)

_____ internal company messengers

_____ messengers provided by an outside service

_____ other (specify) _____

Table C-2 (*continued*)

16. Would you describe your work flow as steady?_____ (Yes/No)

17. Approximately what percent of your job could you delegate to an experienced secretary (work that you do not already delegate)?_____%

18. Please indicate the amount of time (in hours or minutes) you spend each week in each of the following activities:

_____ lunch and breaks
_____ calculating
_____ copying
_____ meetings or discussions in your own office
_____ travel between company locations
_____ telephoning
_____ typing
_____ writing
_____ dictating
_____ proofreading
_____ reading
_____ incoming-mail handling
_____ file handling
_____ other paper handling (sorting, collating, binding, discarding, etc.)
_____ waiting for persons or things
_____ scheduling
_____ searching for information
_____ other (specify) _____

19. What percent of your time do you spend searching for information from the following sources? (The total should equal 100%.)

_____ % in files
_____ % by the phone
_____ % in the library
_____ % in reports or manuals in your office
_____ % other (specify) _____

20. What percent of outgoing documents are distributed by each of the following? (The total should equal 100%)

_____ % hand carrying
_____ % internal mail

Table C-2 (_continued_)

_____ % external mail (postal service)

_____ % TWX® (teletype to teletype transmission)

_____ % facsimile or telecopier

_____ % messenger service

_____ % other (specify) _____

21. For approximately what percent of the documents in question 20 do you need confirmation of the receipt?_____%

22. What percent of the documents that you produce are originated by each of the following? (The total should equal 100%.)

    _____ % dictating to a machine

    _____ % dictating to a secretary

    _____ % writing longhand

    _____ % composing at a keyboard

    _____ % other (specify) _____

23. Of the material you read while in the office, what percent is devoted to each of the following: (The total should equal 100%.)

    _____ % internal mail, within this location

    _____ % internal mail, from other company locations

    _____ % external mail

    _____ % company manuals

    _____ % business periodicals or books

    _____ % files

    _____ % other (specify) _____

24. What percent of the incoming information published by the company do you not need in order to do your job?_____%

25. Does the information in question 24 consist of one or more of the following? (Check the appropriate responses):

    _____ letters or memos

    _____ reports

    _____ manuals

    _____ other (specify) _____

26. How many times do you travel to other cities in an average year? _____

Table C-2 (*continued*)

27. On the average, how many days are you away on each trip? _____

28. What percent of your business travel is devoted to each of the following? (The total should equal 100%.)

   _____ % internal locations nationally, own company
   _____ % internal locations internationally
   _____ % external locations nationally, another company
   _____ % external locations internationally
   _____ % other (specify) _____

29. While traveling, how often do you need information from your office? (Check one.)

   _____ frequently
   _____ sometimes
   _____ seldom
   _____ never

30. How often do you need to access a data base while away from the office? (Check one.)

   _____ frequently
   _____ sometimes
   _____ seldom
   _____ never

31. What percent of your phone calls are placed by each of the following? (The total should equal 100%.)

   _____ % you
   _____ % your secretary
   _____ % other parties (specify) _____

32. Do you answer your own or are your calls screened by a secretary or the switchboard?

   _____ answer own phone
   _____ calls screened

33. What percent of the workday is your phone not covered by anyone?_____%

34. On the average, how many incoming phone calls do you receive in a day? _____

Table C-2 (continued)

35. Of your incoming phone calls, what percent are from each of the following? (The total should equal 100%.)

_____ % people in your own department, this location
_____ % people in other departments, this location
_____ % people at other company locations
_____ % external parties
_____ % international, external
_____ % international, internal

36. On the average how many outgoing phone calls do you place in a day? _____

37. Of your outgoing phone calls, what percent are to each of the following? (The total should equal 100%.)

_____ % people in your own department, this location
_____ % people in other departments, this location
_____ % people in other company locations
_____ % external parties
_____ % international, external
_____ % international, internal

38. If you cannot reach a person, what percent of the time do you do each of the following? (The total should equal 100%.)

_____ % leave a call-back message
_____ % leave a message
_____ % call the person back when you are able to
_____ % go to see the person instead
_____ % other (specify) _____

39. On the average, how many messages are left for you in a day while you are out of the office or on the phone? _____

40. What percent of messages you receive fall into the following categories? (The total should equal 100%.)

_____ % please call
_____ %Will call again
_____ %Returned your call
_____ % a message of substance—something other than "please have him call me"

Table C-2 (*continued*)

41. What is your opinion of recording devices for leaving messages? (Check your response.)

_____ they are useful and I like them

_____ they are useful but I do not like them

_____ they are not useful but I like them

_____ they are not useful and I don't like them

_____ no opinion

42. What percent of your incoming mail is in each of the following categories? (The total should equal 100%.)

_____ % internal, this location

_____ % internal, other company locations

_____ % external

43. What percent of your outgoing mail is in each of the following categories? (The total should equal 100%)

_____ % internal, this location

_____ % internal, other company locations

_____ % external

44. Different people spend relatively different amounts of time in verbal or written communication. What percent of your communication time is: (The total should equal 100%.)

_____ % verbal

_____ % written

45. The mode of communication in question 44 is (Check one.):

_____ completely determined by the job

_____ primarily determined by the job

_____ about equally determined by job and preference

_____ primarily determined by personal preference

_____ completely determined by personal preference

46. If you need to exchange information with someone, your preferred mode of communication would be either of the following: (Check one.)

_____ instant written communication

_____ instant verbal communication

Table C-2 (*continued*)

47. On the average, how many times a month do you or your secretary send/receive documents via facsimile or telecopier? _____

48. Approximately how much time per week do you spend working at home?_____ (hours)

49. What percent of your work at home is in each of the following categories? (The total should equal 100%.)

_____ % reading
_____ % writing
_____ % dictating
_____ % calculating
_____ % telephoning
_____ % typing
_____ % face-to-face meetings, business
_____ % copying
_____ % computer programming
_____ % other (specify) _____

50. What percent of your current in-office tasks could be done at home if you had adequate support equipment?_____%

51. Would you prefer to do this work away from the office if you could?_____ (Yes/No)

52. Do you have a reminder system to check on work in progress, or a "to do" list?_____ (Yes/No)

53. How does your secretary develop an understanding of your work? (Check the appropriate responses.)

_____ she doesn't
_____ written instructions on work
_____ incoming mail
_____ outgoing mail
_____ overhearing your interactions with others
_____ face-to-face interactions with you
_____ other (specify) _____

54. Check the kind of calendar(s) you keep.

_____ month at a glance
_____ week at a glance

Table C-2 (*continued*)

_____ daily

_____ none

_____ other (specify) _____

55. Who maintains your calendar? (Check one.)

_____ I do

_____ my secretary does

_____ my secretary and I do

_____ other (specify)

_____ not applicable

56. Do you carry a calendar with you when you are away from the office?_____ (Yes/No)

57. Approximately what percent of the entries on your calendar does each of the following constitute? (The total should equal 100%.)

_____ % scheduled meetings

_____ % scheduled telephone calls

_____ % deadlines

_____ % reminders

_____ % phone numbers and addresses

_____ % birthdays and nonbusiness events

_____ % other (specify) _____

58. On the average how many entries are contained on your calendar each day? _____

59. Approximately how many times a day do you refer to your calendar? _____

60. What percent of the entries made on your calendar are rescheduled?_____%

61. What percent of the entries scheduled on your calendar are made in each of the following periods? (The total should equal 100%.)

_____ % six months or more in advance

_____ % three months in advance

_____ % one month in advance

_____ % two weeks in advance

Table C-2 (*continued*)

_____ % one week in advance
_____ % one day in advance
_____ % the same day

62. Do you refer to your calendar for information from past entries?
    _____ (Yes/No)

63. Do you maintain files in your office?_____ (Yes/No)

64. What percent of the files you maintain in your office fall in each of the following categories? (The total should equal 100%.)

    _____ % active files, which will eventually be filed elsewhere
    _____ % confidential files kept under lock and key
    _____ % personal files
    _____ % archival files, which could be filed elsewhere
    _____ % files you have pulled from elsewhere and which you will replace or have replaced
    _____ % duplicates of materials that are also filed elsewhere
    _____ % personnel
    _____ % other (specify) _____

65. Why do you need to access files, either in your own office or from your secretary or another area? (Check the appropriate responses.)

    _____ for reference and use in the office
    _____ for reference and to take to other locations
    _____ make hard copy and forward
    _____ update
    _____ discard
    _____ other (specify) _____

66. Are you familiar with your secretary's filing procedure?_____
    (Yes/No)

67. Do you personally pull and use files for which your secretary is responsible?_____ (Yes/No)

68. Who is responsible for replacing files you or your secretary has pulled for your use?

    _____ your secretary
    _____ clerk
    _____ whoever pulled the file

Table C-2 (*continued*)

69. When files shared by several people are pulled, is there a check-out procedure indicating to others the location of these files?

_____ yes

_____ no

_____ don't know

70. How many internally published reference manuals do you routinely use in your work? _____

71. Does your work require calculations?_____ (Yes/No)

72. What percent of your calculations are done by each of the following? (The total should equal 100%.)

_____ % calculator

_____ % adding machine

_____ % paper and pencil

_____ % computer

_____ % other (specify) _____

73. Do you ever calculate while on the phone?_____ (Yes/No)

74. From the time you finish dictating or writing a letter, a report, or other matter, how long (in hours) does it take on the average before the typed version is back on your desk? _____

75. What percent of your letters are in final form on the first typing?_____ %

76. What percent of your written work must you have back on your desk in each of the following times? (The total should equal 100%.)

_____ % less than one hour

_____ % one to three hours

_____ % four to eight hours

_____ % the next day

_____ % two or more days

Table C-2 (*continued*)

77. What percent of your written work would you like to have back on your desk in each of the following times? (The total should equal 100%.)

     _____ % less than one hour
     _____ % one to three hours
     _____ % four to eight hours
     _____ % the next day
     _____ % two or more days

78. What percent of your written work has to be received by the addressee in each of the following times? (The total should equal 100%.)

     _____ % less than one hour
     _____ % one to three hours
     _____ % four to eight hours
     _____ % the next day
     _____ % two or more days

79. What percent of the day is your secretary needed but is not available to you for support tasks?_____%

80. How many times a month do you attend meetings in which you must present information? _____

81. When presenting information, what percent of the time must you use a visual aid (charts, transparencies, foils, slides)?_____%

82. If you use visual aids, what percent of the time do you use each of the following? (The total should equal 100%.)

     _____ % flip charts
     _____ % transparencies/overhead projectors
     _____ % displays on heavy card stock or poster board
     _____ % slide projectors
     _____ % other (specify) _____

83. What percent of your flip chart visual aids are made by each of the following? (The total should equal 100%.)

     _____ % you
     _____ % your secretary/administrative assistant
     _____ % another principal
     _____ % graphic service/outside service
     _____ % other (specify) _____

Table C-2 (*continued*)

84. What percent of your transparencies are made by each of the following? (The total should equal 100%.)

_____ % you

_____ % your secretary/administrative assistant

_____ % another principal

_____ % graphic service/outside service

_____ % other (specify) _____

85. Indicate which office facilities, services, and/or operations listed below are major problem areas that repeatedly interfere with the effective performance of your job.

_____ scheduled meetings in your office

_____ scheduled meetings at other locations

_____ unscheduled meetings

_____ typing service quality

_____ typing service turnaround time

_____ general secretarial services other than typing

_____ dictation equipment and services

_____ facsimile or telecopier service

_____ information retrieval from files by your secretary

_____ information retrieval from files by you

_____ general maintenance of files

_____ incoming phone calls

_____ outgoing phone calls

_____ receiving phone message

_____ number of interruptions

_____ time and facilities for writing

_____ scheduling of activities

_____ time and facilities for calculation or computation

_____ facilities and service for copying

_____ availability and convenience of needed equipment

_____ travel

_____ space

_____ staff support

_____ external business services

_____ internal mail

Table C-2 (*continued*)

86. In the space, below please feel free to comment on your own work style, particular problem areas, and office work in general, or to suggest purchase of equipment which might help you in your job regardless of practicality.

Table C-3   Secretary, Administrative Assistant, and Clerk/typist Questionnaire

Date _____

Name: _____

Title:_____   Your telephone extension _____

I. General

1. Please describe which of your job responsibilities you consider most important.

_____

_____

_____

_____

2. How many people do you work for? _____

3. What are your principals'* names? (i.e., for whom do you work?) and what percent of your work load comes from each of these people (for large groups—for example, programmers, list the job title once and the total work-load percent for the group)?

Name of Principals	Percent Work Load	Name of Principals	Percent Work Load

*A principal is anyone who needs, has, or requests administrative support.

_____

_____

_____

II. Activities

  1. Please indicate the approximate amount of time you spend
(in hours or minutes) *each week* in each of the following
activities in both a normal and a peak-period week:

Normal	*Peak Period*	*Routine Activities*
_____	_____	lunch and breaks
_____	_____	calculating
_____	_____	copying—preparing material for print shop or copying, collating, and distributing copies
_____	_____	face-to-face meetings and conversations
_____	_____	telephone
_____	_____	scheduling meetings/conference/travel other
_____	_____	typing—regular typewriter
_____	_____	typing—automated typewriter
_____	_____	taking dictation
_____	_____	steno
_____	_____	machine
_____	_____	dictating
_____	_____	proofreading
_____	_____	file handling—creating new files, refiling, updating, updating manuals
_____	_____	searching for information (by telephone, in files, in person)
_____	_____	scheduling activities (excluding meetings, conferences, travel)
_____	_____	mail services—incoming (receipt, opening, scanning/sorting, time-stamping, delivery, etc.)
_____	_____	messenger/errand services—calling for messengers, running internal errands, visitor reception

Table C-3 (*continued*)

_____ _____ presentation aids—chart making, preparation of overhead slides, etc.

Administrative Activities (activities that require independent judgment, creativity, and decision making by the secretary)

_____ _____ researching, gathering, assembling, or analyzing information for reports, presentations, or decision making

_____ _____ composing correspondence or responses to correspondence, composing procedures, guidelines, etc.

_____ _____ coordinating or scheduling activities with other principals, outside vendors, outside organizations, etc. (excluding meetings)

_____ _____ answering questions/filling requests for information for items

Miscellaneous Activities

_____ _____ posting/maintaining lists, logs, or reports
_____ _____ preparing input/transmittal forms
_____ _____ document assembly/document control
_____ _____ other record keeping

2. On the average, how much time do you spend each day away from your desk in business-related activities? _____

3. How would you characterize your work load? (Check one.)

_____ heavy volume all the time
_____ normal volume most of the time
_____ peaks and valleys—unpredictable
_____ cyclical—heavy during special projects, months, etc.

4. For peak periods of work activity that occur regularly, please list reasons for the activity (for example, profit plan, principal leaving/returning from trip, quarterly presentations, etc.) and *duration* (for example, 3 days/month, 1 week/quarter, etc.)

_____

_____

_____

_____

5. Do you delegate work to anyone else?_____ (Yes/No) To
   whom? _____

6. Are you responsible for scheduling work for anyone
   else?_____ (Yes/No)

7. Please review the following list of activities and check those
   on which you *receive* assistance or support on a *regular* basis,
   and those on which you *provide* assistance or support on a
   regular basis.

Activity	*Receive Assistance or Support Regularly* (/)	*Provide Assistance or Support Regularly* (/)
Answering telephones		
Making or distributing copies		
Sorting or distributing mail		
Creating, maintaining, purging files		
Scheduling meetings, itineraries		
Typing		
Proofreading		
Creating/preparing/analyzing data or reports		
Creating correspondence, procedures, etc.		
Coordinating/scheduling (excluding meetings)		
Posting/Maintenance Lists and Logs		

Table C-3 (*continued*)

Preparing data entry/maintenance/trans-
mittal forms

(a) Considering your responses to the above-listed
activities, please comment on your feelings about the
arrangements for support (adequate, inadequate
because . . . , etc.).

_____

_____

_____

(b) If clerical assistance were available to you, circle above
or specify below the activities for which you would most
like help (filing, telephone work, setting up meetings,
transcribing dictation, making copies, distributing
copies, distributing mail, other). _____

_____

_____

III. **Equipment/Training**

1. Insert the applicable number in regard to the following
equipment:

1. have access to and use it
2. have access to and do not use it

_____ telephone
_____ mechanical telephone-answering device
_____ intercom
_____ calculator or adding machine
_____ computer terminal
_____ Telex or TWX®
_____ individual system (one with a remote recorder)
_____ individual transcribing unit
_____ facsimile or telecopier
_____ typewriter
_____ automated typewriter/word processing

Table C-3 (*continued*)

2. Have you ever been trained in one or more of the following? (Check the applicable responses.)

 _____ stenography
 _____ computer terminal operation
 _____ computer programming
 _____ automated typewriter operation (for example, Vydec, WANG, IBM)
 _____ other equipment, systems skills, etc. specify: _

IV. Typing

1. What percent of your typing results from the following? (The total should equal 100%.)

 _____ % machine dictation
 _____ % shorthand
 _____ % longhand
 _____ % typewritten copy
 _____ % self-composed
 _____ % other (specify): _____

2. What percent of the typing for your principal(s) is handled by?

 _____ % you
 _____ % other secretaries backing you up
 _____ % word processing center
 _____ % home office word processing center
 _____ % temporary secretaries
 _____ % outside services

V. Copying

1. On the average how many copies of documents that *you* type do you make? _____

2. What percent of your copies are made by each of these methods?

 _____ % carbon paper _____ % copier
 _____ % copysettes _____ % print shop

3. What percent of the documents that you copy do *you* have to collate?_____%

Table C-3 (*continued*)
VI. Telephone

1. How many telephone numbers ring on your phone? _____

2. Please list the extension numbers and the names of the parties.

    Extension   Name   Extension   Name

    _____

    _____

    _____

    _____

    _____

3. Referring to question 2, place an A to the right of the person's name who answers his/her own phone, or an H if you answer it for him/her (in each case when the person is in the office).

4. Approximately how many incoming phone calls do you take in a day? _____

5. Of the incoming phone calls:

    (a) How many can you handle yourself? _____
    (b) How many require messages for your principals? _____
    (c) What percent of messages you take fall into each of the following categories? (The total should equal 100%.)

         _____ % please call
         _____ % will call again
         _____ % a message of substance (caller states the problem, question, etc.)

6. Does anyone cover your phone while you are away from your desk? (Yes/No) _____

Table C-3 (*continued*)

7. Estimate the average amount of time each day that your phone is not covered by anyone. _____

8. On the average, how many outgoing calls do you make in a day? _____

9. About how many of these outgoing calls do you place for your principal(s)? _____

VII. Calculating

1. Does your work require calculations?_____ (Yes/No)

2. What percent of your calculations are done by each of the following? (The total should equal 100%.)

   _____ % calculator
   _____ % adding machine
   _____ % paper and pencil
   _____ % computer
   _____ % other (specify) _____

VIII. Typing

From the time you receive material to be typed or retyped, how long (in hours) does it take on the average before it is back on your principal's desk in typed or retyped form?_____ (Note: One day equals eight hours.)

What percent of the material you are given to be typed is in final form on the first typing(i.e., not edited and put through another draft)_____%

IX. Mail

1. What do you do with incoming mail? (Check appropriate responses.)

   _____ do not handle
   _____ deliver unopened
   _____ open
   _____ time-stamp
   _____ log
   _____ read for information
   _____ highlight

Table C-3 (*continued*)

    _____ attach route slip

    _____ attach relevant files

    _____ assign priority for action

    _____ handle request personally

    _____ answer letter personally

2. Do you perform mail-handling tasks for other people besides your principal(s)?_____Yes/No

3. What kinds of documents that you receive do you read and discard, or read, file, and never use again: _____

X. Manuals, Lists, Logs

1. List the manuals you refer to in your job (a manual is a collection of procedures/guidelines, organized in a specific manner and usually enclosed in a binder—e.g., secretaries' manual, sandards and procedures manual, operating manual): _____

_____.

2. How many copies of manuals are you responsible for keeping up to date? _____

3. Which of the following lists, logs, or quick reference files do you keep? (Check the appropriate responses.)

    _____ telephone numbers

    _____ address lists

    _____ mail log

    _____ equipment records

    _____ other (specify) _____

4. Please describe any problems/comments/suggestions related to maintaining/improving lists (e.g. mailing, employee, customer), logs and other recorded data. _____

_____

_____

XI. Files and Filing

1. What percent of the files that *you* maintain are located at the following places? (The total should equal 100%.)

Table C-3 (*continued*)

_____ % at your desk

_____ % in your immediate work area

_____ % in your principals' offices

_____ % away from your immediate work area

2. Do you maintain files that are shared or used by more than one principal or by other secretaries?_____ Yes/No

3. Has your group developed standard filing systems and procedures that are followed by each person?_____ Yes/No

4. Which of the following applies to the files in your area with regard to reviewing them, discarding them, or sending them to archive storage?

    _____ written guidelines/procedures

    _____ no written guidelines/procedures

    _____ guidelines given verbally by your principal(s)

    _____ self-initiated without guidelines

    _____ files are not reviewed/purged

    _____ do not know

XII. Miscellaneous

1. Indicate which activities, services, and/or facilities listed below consume much of your time, present difficulties in getting your job done, or are areas where you would like to see major improvements.

    _____ scheduling meetings

    _____ scheduling conference or project workrooms

    _____ typing load

    _____ dictation equipment and services

    _____ facsimile service

    _____ information retrieval from files (yours or files of other departments)

    _____ locating information not normally filed in your area

    _____ general maintenance of files

    _____ incoming phone calls

    _____ outgoing phone calls

    _____ taking phone messages

    _____ number of interruptions

    _____ time and facilities for calculation or computation

    _____ facilities and service for copying

Table C-3 (*continued*)

_____ availability and convenience of needed equipment (list equipment needed)

_____ support and cooperation of your principal(s)
_____ support and cooperation of other secretaries in your area or department
_____ mail within the home office
_____ mail to/from branches/laboratories/subsidiaries
_____ manuals (preparing, maintaining, distributing, etc.)

2. Briefly describe below projects, activities, or responsibilities that you would like to take on or initiate, and give a brief description of the benefits that would result.

Insert an asterisk following those projects which you feel require the efforts of more than one secretary, administrative assistant, or clerk/typist to plan and implement.

3. Please make any other comments or suggestions you wish.

Table C-3 (*continued*)

_____

_____

_____

_____

_____

_____

Thank you for your cooperation.

Table C-4   Secretary Questionnaire

1. Name: _____

2. Title. _____

3. Department: _____

4. Location: _____

5. Work phone number and extension: _____

6. How many people do you work for? _____

7. What are your principals'* names? (i.e., for whom do you work?) _

   _____

   _____

8. What percent of your work load comes from each of these people?

   _____

   _____

   _____

9. Do you do supervisory or coordinating work for other people?_____ (Yes/No)

10. Do you delegate work to other secretaries?_____ (Yes/No)

11. Are you responsible for scheduling work for other secretaries?_____ (Yes/No)

12. Number of years as a secretary: _____

13. Number of years in this company: _____

*A principal is anyone who needs, has, or requests administrative support.

Table C-4 (*continued*)

14. Number of years in this position: _____

15. Have you been trained in one or more of the following? (Check applicable responses.)

    _____ stenography
    _____ transcription from machines
    _____ computer terminal operation
    _____ keypunch
    _____ computer programming
    _____ magnetic typewriter operation
    _____ facsimile or telecopier operation
    _____ other (specify) _____

16. Fill in the blanks with the appropriate letters in regard to the following equipment:

    a. located in my own office and I use it
    b. located in my office but I do not use it
    c. have access to and use it
    d. have access to and do not use it
    e. equipment not available

       _____ telephone
       _____ mechanical telephone-answering device
       _____ speaker phone
       _____ intercom
       _____ calculator or adding machine
       _____ computer terminal
       _____ Telex or TWX®
       _____ individual dictation unit
       _____ dictation system (one with a remote recorder)
       _____ transcriber
       _____ facsimile or telecopier
       _____ copier
       _____ typewriter

17. How many telephone numbers ring on your phone? _____

18. Check the kind of typewriter you use:

       _____ video display or CRT typewriter
       _____ magnetic card, tape, or memory

Table C-4 (*continued*)

_____ manual, electric, or correcting
_____ other (specify) _____
_____ none

19. Would you describe your work flow as steady?_____ (Yes/No)

20. Please indicate the approximate amount of time you spend (in hours or minutes) each week in each of the following activities:

_____ lunch and breaks
_____ calculating
_____ copying
_____ face-to-face meetings and conversations
_____ telephoning
_____ typing
_____ writing
_____ record keeping
_____ dictating
_____ taking direction
_____ proofreading
_____ reading
_____ incoming-mail handling
_____ file handling
_____ other paper handling (sorting, collating, binding, discarding, etc.)
_____ waiting for persons or things
_____ scheduling
_____ searching for information
_____ other (specify) _____

21. Of the time you spend searching for information, on the average what percent is devoted to searching in each of the following ways (The total should equal 100%.)

_____ % in files
_____ % by the phone
_____ % in the library
_____ % in reports or manuals in your office
_____ % other (specify) _____

22. What percent of outgoing documents are distributed in each of the following ways? (The total should equal 100%.)

Table C-4 (*continued*)

_____ % hand-carrying them

_____ % internal mail

_____ % external mail (postal service)

_____ % TWX® (teletype to teletype transmission)

_____ % facsimile or telecopier

_____ % messenger service

_____ % other (specify) _____

23. If you use messenger services, approximately what percent of the time do you use each of the following? (The total should equal 100%.)

_____ % internal company messengers

_____ % messengers provided by an outside service

_____ % other (specify) _____

24. In approximately what percent of the services in question 23 do you need confirmation of the receipt?_____%

25. On the average what percent of your typing is done from each of the following? (The total should equal 100%.)

_____ % machine dictation

_____ % shorthand

_____ % longhand

_____ % typewritten copy

_____ % self-composed

_____ % other (specify) _____

26. How many times a month on the average do you send/receive documents via facsimile or telecopier? _____

27. Do you have a typewriter at home?_____ (Yes/No)

28. How many hours a week do you spend typing at home? _____

29. List the specific kinds of typing you do at home: _____

_____

_____

Table C-4 (continued)

30. Do you have a reminder system (follow-up or tickler) to check on work in progress, or a "to do" list?_____ (Yes/No)

31. How do you develop an understanding of your principal(s) work? (Check the appropriate response or responses.)

    _____ written instructions on work
    _____ incoming mail
    _____ outgoing mail
    _____ overhearing your principal's interaction with others
    _____ face-to-face interaction with your principal
    _____ other (specify) _____

32. How does your principal(s) give your instructions for your work? (Check the appropriate response.)

    _____ gives them in person
    _____ writes them in longhand
    _____ calls on the phone
    _____ calls on intercom
    _____ dictates them
    _____ other (specify) _____

33. What activities do you perform with incoming mail? (Check the appropriate responses.)

    _____ do not handle
    _____ deliver unopened
    _____ open
    _____ time-stamp
    _____ log
    _____ read for information
    _____ highlight
    _____ attach routing slip
    _____ attach relevant files
    _____ assign priority for action
    _____ handle request personally
    _____ answer letter personally

34. What percent of the documents you receive do you handle in each of the following ways? (The total should equal 100%.)

    _____ % read and discard
    _____ % read, file, and never use again

Table C-4 (*continued*)

35. On the average, how many pieces of incoming mail do you handle a day, excluding magazines, brochures, advertisements, newspapers, etc.? _____

36. What percent of the incoming mail consists of each of the following? (The total should equal 100%.)

    _____ % internal, this location
    _____ % internal, other company locations
    _____ % external, another company

37. On the average, how many pieces of outgoing mail do you send in a day? _____

38. What percent of your outgoing mail consists of the following? (The total should equal 100%.)

    _____ % in-house
    _____ % external
    _____ % within department

39. How much working time on an average day do you spend away from your desk?_____hours

40. Which of the activities listed below do you perform while away from your desk? (Please check those items that apply.)

    _____ copying
    _____ TWX®
    _____ facsimile or telecopier
    _____ meetings or conversations
    _____ reception
    _____ errands
    _____ libraries
    _____ searching for people
    _____ getting supplies
    _____ visitor escort
    _____ training sessions
    _____ cashier
    _____ picking up and delivering mail
    _____ covering someone else's desk
    _____ telephone switchboard
    _____ other (specify) _____

41. What percent of your principal(s) calls do you place?_____%

Table C-4 (*continued*)

42. What percent of the workday is your phone not covered by anyone?_____%

43. About how many incoming phone calls do you take in a day? _____

44. Of incoming phone calls, what percent are from each of the following? (The total should equal 100%.)

    _____ % people in your own department, this location
    _____ % people in other departments, this location
    _____ % people at other company locations
    _____ % external parties
    _____ % international, external
    _____ % international, internal

45. About how many outgoing phone calls do you place in a day? _____

46. Of the outgoing phone calls, what percent are to each of the following? (The total should equal 100%.)

    _____ % people in your own department, this location
    _____ % people in other departments, this location
    _____ % people in other company locations
    _____ % external parties
    _____ % international, external
    _____ % international, internal

47. On the average, how many messages do you take for your principal(s) in one day while he or she is (they are) out of the office or on the telephone? _____

48. What percent of the messages you take fall into each of the following categories? (The total should equal 100%.)

    _____ % please call
    _____ % will call again
    _____ % returned your call
    _____ % a message of substance—something other than "please have him call me"

49. What is your opinion of recording devices for leaving messages? (Check your response.)

    _____ they are useful and I like them
    _____ they are useful but I do not like them
    _____ they are not useful but I like them

Table C-4 (*continued*)

_____ they are not useful and I do not like them

_____ no opinion

50. Different people spend relatively different amounts of time in verbal or written modes of operation. What percent of your activity time is devoted to each of the following? (The total should equal 100%.)

_____ verbal

_____ written

51. The mode of operation in 50 is: (Check one.)

_____ completely determined by the job

_____ primarily determined by the job

_____ about equally determined by job and personal preference

_____ primarily determined by personal preference

_____ completely determined by personal preference

52. If you need to exchange information with someone, your preferred mode of communication would be one of the following.

_____ % personal

_____ % phone

_____ % video conference

_____ % in writing

_____ % other (explain) _____

53. Check the kind of calendar(s) that you maintain.

_____ month at a glance

_____ week at a glance

_____ daily

_____ none

_____ other (specify) _____

54. Do you keep a duplicate of your principal(s) calendar?_____ (Yes/No)

55. Approximately what percent of the entries on your calendar consist of each of the following? (The total should equal 100%.)

_____ scheduled meetings

_____ scheduled telephone calls

_____ deadlines

_____ reminders

_____ phone numbers and addresses

Table C-4 (*continued*)

_____ birthdays and nonbusiness events

_____ other (specify) _____

56. On the average, how many entries are on the calendar each day? _

57. How many times a day do you refer to the calendar? _____

58. What percent of the entries scheduled on the calendar are made in each of the following periods? (The total should equal 100%.)

_____ % six months or more in advance
_____ % three months in advance
_____ % one month in advance
_____ % two weeks in advance
_____ % one week in advance
_____ % one day in advance
_____ % the same day

59. Do you refer to the calendar for information from past entries?_____ (Yes/No)

60. How many internally published reference manuals do you routinely use in your work? _____

61. Which of the following lists, logs, or quick reference files do you keep? (Check appropriate responses.)

_____ telephone numbers
_____ address lists
_____ mail log
_____ equipment records
_____ other (specify) _____

62. Do you maintain files in your office?_____ (Yes/No)

63. Do you use a standard filing procedure?_____ (Yes/No)

64. What procedure(s) is (are) used to clean out (purge—send files to storage or throw them away) files? (Check appropriate responses.)

_____ standard procedure
_____ guidelines of the people you assist
_____ self-initiated procedure
_____ never purge files
_____ do not know

Table C-4 (continued)

65. Is your principal(s) familiar with your filing procedure?_____
(Yes/No)

66. Do you personally pull and use the files for which other secretaries
are responsible?_____ (Yes/No)

67. Who is responsible for replacing files that have been pulled?

68. For files that are shared by several people, is there a check-out
procedure whereby others can locate the files if they have been
pulled?

_____ yes
_____ no
_____ do not know

69. Does your work require calculations?_____ (Yes/No)

70. What percent of your calculations are done by the following? (The
total should equal 100%.)

_____ % hand
_____ % calculator
_____ % computer

71. Do you ever do calculations while you are on the phone?_____
(Yes/No)

72. From the time you receive a dictated or drafted letter or other
matter, on the average how long (in hours) does it take before it is
back on your principal's desk in typed form?_____ (Note: One
day equals eight hours.)

73. What percent of the letters are in final form on the first typing
(i.e., are not edited and put through another draft)?% _____

74. Indicate which office facilities, services, and/or operations listed
below are major problem areas that repeatedly interfere with the
effective performance of your job.

_____ scheduling meetings in your principal's office
_____ scheduling meetings at other locations
_____ unscheduled meetings
_____ typing load
_____ dictation equipment and services
_____ facsimile or telecopier service
_____ information or retrieval from files

Table C-4 (continued)

_____ general maintenance of files

_____ incoming phone calls

_____ outgoing phone calls

_____ taking phone messages

_____ number of interruptions

_____ time and facilities for reading

_____ time and facilities for writing

_____ time and facilities for calculation or computation

_____ facilities and service for copying

_____ availability and convenience of needed equipment

_____ staff support

_____ internal mail

75. In the space below please feel free to comment on your own work style, particular problem areas, office work in general, or make suggestions for new processes and work patterns that might help you in your job regardless of practicality.

_____

_____

_____

_____

_____

_____

_____

_____

_____

Sample Outline for Request for Proposal and Sample Questionnaire for Vendor Selection

REQUEST FOR PROPOSAL (RFP) INSTRUCTIONS
(Please read before proceeding)

1. **Questionnaires**
   (a) All questions must be answered. Responses from companies who submit *incomplete* questionnaires *will not be accepted.*
   (b) *Answer* all questions directly on the questionnaire. Responses must be typewritten. Attach separate sheet(s) if additional space is required to complete answers.

2. **Quotations:** On the basis of software and/or hardware equipment and of services, submit quotations for:
   Purchase terms (including maintenance terms)
   Rental terms (including maintenance terms)
   Lease terms (including maintenance terms)

3. **Proposals:** Submit in narrative or outline form; also submit product brochures or documents highlighting the applicable points in support of the vendor proposal.

4. **Due Date:** Responses received after this date will not be considered.

5. **Expiration Date:** Indicate expiration date of proposal.

6. **Authorized Signature:** Questionnaires, quotations, and proposals must be signed by an authorized executive in the vendor firm. Improperly authorized responses *will not be accepted.*

7. **Return:** Questionnaires, quotations, and proposals should be returned to:
   Name of Company
   (Attention of) Name of Individual
   Address

QUESTIONNAIRE FOR VENDOR SELECTION

Company Profile

VENDOR COMPANY NAME: _____

PARENT COMPANY NAME: _____

288

PRODUCTS: _____

_____

_____

_____

_____

_____

_____

SERVICES: _____

_____

_____

_____

AUTHORIZED VENDOR SIGNATURE: ____ DATE: _____
TITLE: _____

1. Attach a copy of your company's latest annual report. How many years has your company been in business? _____
2. Indicate corporate and other division, subsidiary or license affiliations.
3. Attach appropriate financial statements reflecting profitability.
4. For each product under consideration for purposes of this quotation, indicate your:
   (a) Current market share
   (b) Date of introduction
   (c) Number of units sold since the year
   (d) Attach a list of references (minimum of four)
5. Major products and services.
6. What percentage of your sales are software?
7. What percentage of your sales are consulting services?

8. What percentage of your sales are custom software (programming)?

9. Future products under development (we will consider a nondisclosure agreement for this purpose).

10. Home office location

11. International office locations

12. U.S. locations

13. Describe services provided and typical project arrangements with your customer or clients.

14. Describe primary business function.

15. Does each office location operate as an autonomous or independent profit center? (Please explain.)

16. Are managers in each location independent in terms of managing their territories and business decision trade-offs?

17. Are your products and/or services centrally coordinated or managed from a corporate office within the United States and worldwide?

18. Do you offer price discounts or a sliding scale for products and/or services?

19. Which office in your company (and the title of the person) would be a worldwide focal point? What is the normal arrangement for the coordinator? Is this support available for a cost *only?*

20. What would the support be between your area support office and the other domestic and/or international locations?

21. *Support:*
    (a) What is your response time from the time a service call is placed?
    Average time _____ Maximum time _____
    (b) Please comment on the availability of hardware/software failure guarantees and backup service?
    Will software/hardware fixes be made within 24 hours (yes/no)?_____
    Comments_____

Training and Support

22. Training and applications support:
    (a) How long is the initial formal operator-training period for each of your products under considerations?
    Computer operator _____ Terminal users _____
    (b) Where is training available?
    (c) How many operators and professionals are trained free with each system?

		Number Trained Free with system	Cost to train additional operators
	Computer	_____	_____
	Terminal	_____	_____
	Other	_____	_____

(d) How long will it take an operator to reach proficiency after formal training?

		Reasonable Operation	Peak Proficiency
	Computer	_____	_____
	Terminal	_____	_____
	Other	_____	_____

(e) Will an applications support representative be available:
During installation (yes/no)? ___ How long? ____ Cost ____
Follow-up support (yes/no)? ___ How long? ____ Cost ____

(f) Is programming support provided by your company?
(yes/no)_____ If yes, please explain:

23. Does training include management orientations (yes/no)_____
    a) Is there a fee?_____yes/no
    b) Please comment on the general level of the orientation (e.g. technical, financial): _____

Contracts and Quotations

24. *Contracts*
    (a) Please attach a copy of all purchase, rental, lease, and service agreements.
    (b) Is a national account program available? (yes/no) _____
    (c) What quantity discounts are available? _____
    (d) What rental plans are available? _____
    (e) What lease plans are available? _____
    (f) What cancellation provisions are provided in purchase, rental, and lease agreements? _____
    (g) Would a clause be included in the contract stating cancellation of the contract if we are not satisfied with performance, of which we will be the sole judge? (yes/no)_____Explain: _____
    (h) Would an on-site free trial test period of hardware and software at the location of our choice for 90 days be acceptable? (yes/no) Explain: _____
    (i) Once the system is installed, is reimbursement for downtime exceeding four hours applicable? (yes/no)_____. If no, when does reimbursement start?

25. *Quotation*
    (a) Please submit a formal quotation for each product in your product line applicable to the requirement.
    (b) In addition, please summarize your figures below (please include all detail in your quotation):

Hardware/ Software	*Purchase ($)*	*3-year Lease*	*3-year Rental*	*Annual Maintenance*
Hardware/ Software	_____	_____	_____	_____
Other	_____	_____	_____	_____

Software/Communications/Hardware

26. Indicate compilers supported (e.g., Basic, Fortran, Pascal, Cobol, RPG).
27. Explain type and function of operating system. _____
    (a) Indicate or attach list of utilities? _____
28. Is system user-programmable?_____. If so, explain (e.g.: via compilers, macros, procedural language, or unique system language)? _____
29. Is a data-base management system available?_____ (yes/no) Explain capabilities: _____
30. Data Storage
    (a) Total on-line storage _____
    (b) Amount required for operating system _____
    (c) Net amount of storage required for data and applications programs _____
    (d) Access methods supported (ISAM, VSAM, etc) _____
    (e) Number of user files supported _____
    (f) Maximum length of a data record _____
    (g) Maximum number of records per file _____
    (h) Amount of storage (in bytes) required for 1000 characters. ___
    (i) Are variable-length records supported? _____
    (j) Indicate file integrity and recovery procedures. _____
    _____
    _____
31. File Management _____
    (a) Explain indexing capability. _____
    (b) Does index have "tiered" or hierarchical capability (see requirements)? _____
    (c) What status information is available on index? _____
    (d) On what index fields can a file be accessed? _____
    (e) Password security? (yes/no)_____Explain levels: _____

(f) Indicate response times for file access and filing with
simultaneous users as specified.
Local: ＿＿＿ 5 users; ＿＿＿ 20 users; ＿＿＿ 32 users
Remote: ＿＿＿ 5 users; ＿＿＿ 20 users; ＿＿＿ 32 users ＿＿

(g) Other file management functions. ＿＿＿＿＿＿＿＿＿＿

＿＿＿＿＿＿＿＿＿＿＿＿＿＿＿＿＿＿＿＿＿＿＿

32. Communications
(a) Do you support terminals *on line?* (yes/no)＿＿＿＿. List makes
and models and description of terminal functions.
(b) What communications protocols and speeds are supported?
(c) Which protocols run in background and which run in
foreground?
(d) What other vendor equipment does your equipment interface
with?
(e) Do you support any domestic or international
communications protocols?＿＿＿＿(yes/no)
If yes, list them: ＿＿＿＿＿＿＿＿＿＿＿＿＿＿＿
(f) Do you support or interfere with local area networks
(LAN)?＿＿＿＿(yes/no)
If yes, please list them: ＿＿＿＿＿＿＿＿＿＿＿＿＿

＿＿＿＿＿＿＿＿＿＿＿＿＿＿＿＿＿＿＿＿＿＿＿

33. Hardware
(a) Please explain architecture employed.
(b) Indicate CPU and memory sizes.
(c) Indicate type and capacity of disk storage.
(d) Indicate peripherals supported: ＿＿＿＿＿＿＿＿＿＿
(e) Attach hardware system configuration proposed with software.

34. Systems Integration Architecture
(a) Describe all office systems products and your integration
strategy within your product lines and with other vendors.
(b) Indicate industry or company conventions followed or
adopted.
(c) If interactions/interface with your own products or other
vendor products is not yet available, indicate objectives,
potential timing, and availability.

Sample Logs—Mail, Copier, and Mail-Related Costs and Volumes

Table E-1 Incoming Mail Log

Name of Person Receiving Incoming-Mail Item	Document Type*	Number of People Mail Addressed To	Number of People on CC List	Date on Mail Item	Total Pages of Mail Item	Type In		
						Your Department	Other Departments	Other Locations

*L = Letter   B = Procedures Bulletin/Manual
M = Memo   LI = List
P = Personal Mail   F = Form

Table E-1 (*continued*)

Name			
Date	Department		Floor

Indicate Number of Pages of Mail			Mail Coming From			Routed Mail	
Drawn Chart or Graph	Pic- ture	Com- puter Report	Your Depart- ment	Other Home Office Depart- ments	Other Loca- tions	Has Routing Slip Attached	Has Hand Written Notes On It

Number Mail Items Rerouted Due to Incorrect Delivery

Table E-2 Copier Log

NAME: _____
DEPARTMENT: _____
DATE: _____

Document Description	Document Type*	Original Copy Size					# of Originals	# of Copies Per Original
		$8\frac{1}{2} \times 5\frac{1}{2}$	$8\frac{1}{2} \times 11$	$8\frac{1}{2} \times 14$	11×17	Other		

Document Description	Document Type*	Original Copy Size					# of Originals	# of Copies Per Original
		8½ × 5½	8½ × 11	8½ × 14	11 × 17	Other		

*Document Types

L	= Letter	M	= Magazine
M	= Memo	B	= Book
CR	= Computer Report	N	= Newspaper
TR	= Typed Report	LI	= List
F	= Form	O	= Other

Table E-3 Summary of Mail-related Costs and Volumes

Major Operating Unit _____

Prepared By _____ Date _____

Mail Operations -1	Total Number of Full-Time Employees Supported	Number of Activities Reported 2	Estimated Mail Volume (In Number of Pieces)-3					Total Postage Costs (Include All Types)-4	Total Other Mail-Related Costs (All Except Postage)-5
			Incoming		Outgoing		Total		
			Internal	External	Internal	External			
Mail service activities (company mailrooms)									
Other departments involved in mail operations									
Outside direct mail operations									
Shipping and receiving functions									
Product mailing activities (If not included in shipping and receiving)									
Others (specify)									
Totals for major operating unit									

Labor costs (salaries and fringes)-6

Total outside (noncompany employee) messenger costs-7

Total motor vehicle costs related to mail handling-8

Bibliography

BOOKS

Ackoff, Russell L., *A Concept of Corporate Planning* (New York: Wiley, 1970).

Alford, C. P. and J. R. Barges, *Production Handbook* (New York: Ronald Press, 1950).

Anthony, R. N., *Planning and Control Systems* (Boston: Graduate School of Business Administration, Harvard University, 1965).

Bennis, Warren C., Kenneth D. Benne, and Robert Chin, editors, *The Planning of Change*, 3rd ed. (New York: Holt, Rinehart and Winston, 1967).

Brandon, Dick H., *Data Processing Cost Reduction and Control* (New York: Van Nostrand–Reinhold, 1978).

Buffa, E. S., *Modern Production Management* (New York: Wiley, 1969).

Chandler, Alfred D., *Strategy and Structure* (Cambridge: M.I.T. Press, 1962).

Cremach, H.P., *Work Study in the Office* (London: Macolaren & Sons, 1969).

Drucker, Peter F., *Management-Tasks, Responsibilities, and Practices* (New York: Harper & Row, 1974).

Drucker, Peter F., *Managing for Results* (New York: Harper & Row, 1964).

Flaherty, John E., *Managing Change* (New York: Heller Publishing Company, 1979).

Galbraith, Jay R., *Organization Design* (Reading, Mass.: Addison-Wesley, 1977).

Gillbreth, Frank B., *Motion Study* (New York: Van Nostrand–Reinhold, 1911).

Hall, Richard, *Organizations: Structure and Process* (Englewood Cliffs, N.J.: Prentice-Hall, 1978).

Head, Robert V., *Strategic Planning Systems for Information Systems* (Wellesley, Mass.: Q.E.D. Information Sciences, 1979).

Hicks, Herbert G. and James D. Powell, *Management, Organization and Human Resources* (New York: McGraw-Hill, 1976).

Horton, Forest W., Jr., *Information Resources Management* (Cleveland, Ohio: Association for Systems Management, 1979).

Knight, Kenneth, *Matrix Management* (New York: Gower Press, 1971).

Krick, E. V., *Methods Engineering* (New York: Wiley, 1962).

Martin, James, *Future Developments in Telecommunications* (Englewood Cliffs, N.J.: Prentice-Hall, 1979).

Martin, James, *Teleprocessing Network Organization* (Englewood Cliffs, N.J.: Prentice-Hall, 1970).

Martin, James, *The Wired Society* (Englewood Cliffs, N.J.: Prentice-Hall, 1978).

Mintzberg, Henry, *The Nature of Managerial Work* (New York: Harper & Row, 1973).

McLean, Ephraim R. and John V. Soden, *Strategic Planning for MIS* (New York: Wiley, 1977).

Newman, Derek, *Organization Design* (Philadelphia Pa: International Ideas, 1973).

Taylor, Frederick W., *Shop Management* (New York: Harper & Row, 1911).

Toffler, Alvin, *Future Shock* (New York: Random House, 1970).

Rostow, W. W., *The Stages of Economic Growth*, 2nd ed. (Cambridge: University Press, 1971).

Whitmore, *Measurement and Control of Indirect Workers* (New York: American Elsevier, 1979).

PERIODICALS, REPORTS AND OTHER SOURCES

Bair, James H., "Communications in the Office of the Future: Where the Real Payoff May Be," *Business Communications Review,* January–February 1979, Vol. 9, No. 1.

Bair, James H., *Productivity Assessment of Office Automation Systems,* Vols. I and II, Menlo Park, Calif.: SRI International (Prepared for National Archive and Records Service, Office Records Management, Washington, D.C.), March 1979.

Baxter, Robert I. and George F. Krall, "Six Stages Toward the Automated Office," *Words,* October–November 1979, pp. 20–24.

Booz, Allen & Hamilton, Inc., *Why Automate,* Special report, 1980.

"Bubble Memory Devices to Play Increasing Role," *Communications News,* July 1979, p. 83.

Burns, J. Christopher, "The Evolution of Office Informations Systems," *Datamation,* April 1977.

Canning, Richard C., "The Automated Office—Part I," *EDP Analyzer,* September 1978, Vol. 16, No. 9.

Canning, Richard C., "The Automated Office—Part II," *EDP Analyzer,* October 1978, Vol. 16, No. 10.

Connell, John J. "The Office of the Future," *Journal of Systems Management,* February 1979, pp. 6–10.

Connell, John J., *Office of the 80's—Productivity Impact,* Office Technology Research Group, Pasadena, Calif., Special Advertisement, 1980.

Conrath, David W., "Measuring the Impact of Office Automation Technology Needs, Methods and Consequences," *Proceeding of Office Automation Conference,* Carmel, Calif.: Stanford University, 1980.

Davis, George, "AT&T Answers 15 Questions About Its Planned Service," *Data Communications,* February, 1979, pp. 41–60.

Diebold, John, "IRM: New Directions in Management," *Infosystems,* October 1979, pp. 41–43.

Drageset, Dan, "Users of Office Automation," paper presented at AIIE Office Automation Conference, New York, 1978.

Driscoll, James W., *Office Automation: The Organizational Redesign of Office Work,* Cambridge: M.I.T. Working Paper No. 116476, May 24, 1979.

Gantz, John, "The Secret and Promises of Fiber Optics," *Computerworld,* 1980, pp. 1–10.

Hammer, Michael, *Why Is an Office?* Massachusetts Institute of Technology, Cambridge, MA. (October 22–24, 1979), Diebold Research Program Presentation.

Hammer, Michael and Marvin Sirbu, *What Is the Automated Office?* paper presented at 1980 National Computer Conference on Office Automation Conference, Massachusetts Institute of Technology, Cambridge, 1980.

International Data Corporation, "Information Processing and the Office of Tomorrow," *Fortune,* October 1977.

International Data Corporation Special Report, "Productivity and Information Systems for Tomorrow's Office, *Fortune,* September 1980.

Kleper, Michael, "Everything You Always Wanted to Know About In-Plant Phototypesetting in Less Than One Hour," Bedford, Mass: Compugraphic Corporation, 1978.

Nolan, Richard L., "Managing the Crises in Data Processing," *Harvard Business Review,* March–April 1979, pp. 115–126.

Poppel, Harvey L., "The Automated Office Moves In," *Datamation,* August 1979, pp. 73–79.

Rhodes Wayne L., Jr., "Facsimile—New Life for an Old Idea," *Infosystems,* September 1979, pp. 42–52.

Rockart, John F., "Chief Executives Define Their Own Data Needs," *Harvard Business Review,* March–April 1979, pp. 81–92.

Scientific American, "Microelectronics," September 1977. (series of articles)

Selig, Gad J., "Strategic Planning for the Information Systems Resources Functions in a Multinational Environment," Doctoral Dissertation, Pace University, New York, 1980.

Stanford Research Institute, *Office of the Future,* Guidelines, No. 1001, April 1976.

Strassmann, P.A., "Stages of Growth," *Datamation,* Vol. 22, No. 10, October 1976.

Tellefson, Gerald, "Productivity Revisited," in *Summary of Proceedings,* Office Technology Research Group, Pasadena, Calif., September 1980.

"The Evolution of Office Information Systems," *Datamation,* April 1977. (series of articles)

"The Office of the Future," *Business Week,* June 30, 1975.

Tomio, Wada, "Liquid Crystal Research Updates LCD Technology," *Journal of Electrical Engineering,* January 1979, pp. 38–41.

"Turning Telephones into Terminals," *Business Week,* October 1, 1979, pp. 86–90.

White, Robert B., "A Prototype for the Automated Office," *Datamation,* April 1977.

Withington, Frederick G., "Transformation of the Information Industries," *Datamation,* November 1978, pp. 8–14.

Zisman, M. D., *Office Automation; Revolution or Evolution,* Sloan Management Review, Spring, 1978, Vol. 19, No. 3.

Zubel, N. S., R. L. Frank, and A. M. Mayfield, "The Emerging Real World of Office Automation," Arthur D. Little Report, June 1979.

Glossary

Address A coded representation of the destination of data, or their originating point.

Administrative costs Cost related to indirect versus direct labor processes. They include such items as labor, rent, utilities, postage, travel. They are generally chargeable to managerial, sales, and associated administrative support processes.

Administrative support Typically refers to all nontyping and typing secretarial and clerical support functions such as filing and making appointments and travel reservations.

Algorithm A method used for performing a task.

Alphanumeric Pertaining to a character set that contains letters, digits, and usually other characters, such as punctuation marks.

Analog transmission A technique for transmitting information as a continuous signal, by raising and lowering the frequency.

Analysis The methodical investigation of a problem and the separation of the problem into smaller related units for further detailed study.

Analyst A person who defines problems and develops algorithms and procedures for their solution.

Application program (software) A program written for or by a user that applies to a particular application.

Architecture The fundamental principles and design behind a hardware or software product.

Archives Storage facilities for corporate records which are retained for future reference relative to regulatory, tax, financial or other purposes.

Asynchronous Without regular time relationship; unexpected or unpredictable with respect to the execution of a program's instructions.

Auto-Answer Pertaining to a machine feature that allows a transmission control unit or a station to automatically respond to a call that it receives over a switched line.

Auto-Call Pertaining to a machine feature that allows a transmis-

306

sion control unit or a station to initiate a call automatically over a switched line.

Automated teller systems Area known as automated teller machines (ATM). Commonly provided by banking institutions to provide cash disbursements to customers. Area utilized by other types of corporations (American Express) to disperse travelers' checks and cash.

Automatic document storage and retrieval Computer controlled information storage and retrieval. The information may be stored either electronically or in microimage form.

Basic A programming language with a small number of commands and a simple syntax, primarily designed for numerical applications.

Baud A unit of signaled speed equal to the number of discrete conditions of signal events per second.

Bi-directional The ability of a printer mechanism to move left to right and right to left along the print line to speed up the printing process.

Bit Smallest unit of information in the form of a binary digit.

"Black box" Hardware and/or software used to interface unlike systems so that information can be transferred and integrated from one device to another.

Boilerplate Constant information stored electronically that will be reused as part of other documents (e.g., the main body of a form letter).

Boldface To overstrike or offset a printed character resulting in a darker and larger character than normal.

bps Bits per second. In serial transmission, the instantaneous bit speed with which a device or channel transmits a character.

Budgets The planning and establishment of a corporation's expenses, capital needs, and income for some established time frame (usually monthly, quarterly, and annually).

Bug A mistake or malfunction.

Business communications system A complete communications system for the office. The components include; creation, capture, keyboarding, expansion, distribution, storage and retrieval, and disposal of all forms of information.

Business system components   Individual categories of the major business components including: creation, capture, keyboarding, distribution, storage, retrieval, and disposal that can be applied to the processes of all worker categories.

Cache buffer   A temporary, very high-speed memory that some computers use to store the most often accessed data and instructions. Synonyms: Cache, Highspeed buffer.

Capture   Business system component, which includes technologies such as dictation, word processing, optical character recognition, source document automation, and digital voice computer resources.

Career pathing   The establishment or existence of a line of progression from one job position to another, ensuring job opportunities and personal growth.

Cathode ray tube (CRT)   An electronic vacuum tube, such as a television picture tube, that can be used to display information.

Centralized organization   An organization that has a strong central management policy and philosophy established. Generally, divisions have little latitude and freedom in this structure.

Character   A letter, digit, or other symbol used as part of the organization, control, or representation of information.

Characters per inch (CPI)   A unit of measure representing the number of characters in a horizontal inch.

Characters per second (CPS)   A unit of measure representing the number of characters printed in one second by a printer.

Character recognition   The identification of characters by automatic means.

Charter   A statement that outlines the scope and purpose of a department.

Chip   A very small package (usually less than one inch square) containing powerful electronic circuitry. Synonym: Integrated Circuit.

Circuit   A means of two-way communications.

Coaxial cable   A communications line, insulated from external interference and capable of carrying many transmission signals simultaneously.

COBOL   (Common business-oriented language.) An English-like programming language designed for business data processing applications.

Communication   An exchange of information between two or more people.

Compatibility   The ability to move programs and data from one system to another without changing either the program or the information by either communications or media transfer.

Computer (CPU)   A data processor (or central processor unit) that can perform substantial computation, including numerous arithmetic or logic operations.

Computer input microfilming   The transformation of microimages into computer readable information.

Computer output microfilming   The transformation of computer readable information into microimages.

Computer program   A series of instructions or statements in a form acceptable to a computer.

Configuration   The group of links, nodes, machine features, devices, and programs that make up a data or word processor system or a communications network.

Control   The management process of monitoring progress against a plan or standard.

Cost of money   The cost to a corporation for its capital, usually expressed as a percentage.

Creation   The act of thinking and formulating a communication.

Critical path methodology (CPM)   A scheduling technique that allows a project manager to determine the most critical tasks necessary for project completion and their impact on all other tasks.

Cursor A symbol on a display device that can be moved to indicate a character to be changed, added or deleted.

Daisywheel Three-inch diameter disc with spokes radiating from the center. Each spoke contains a character for printing.

Data A presentation of characters in different forms to form meaningful information.

Data base Interrelated data stored together for use by one or more applications. The data are stored so that they are independent of the applications.

Data base management system (DBMS) Software designed to execute the data manipulation for application programs using a data base.

Data management The function of controlling the acquisition, analysis, storage, retrieval, and distribution of information.

Decentralized organization An organization where the business units have a great deal of latitude and freedom from the corporate office.

Decision support systems A computer information system designed for and by managers and professionals using simple interactive programming commands with access to one or more data bases.

Delegable activities Those assignments that can be made to a lower level managerial, professional, technical, secretarial, or clerical employee by a superior. In the office automation environment, these are due to the results of productivity enhancements based on system implementation.

Demographics The study and analysis of information concerning populations on the basis of different variables such as geography, income, and age.

Digital transmission A technique for transmitting information as discrete pulses representing encoded bits of information.

Direct impression A printing technique whereby each character strikes the paper, such as conventional typewriter or impact printers.

Disk storage A magnetic storage medium in which information is stored by magnetic recording on the flat surfaces of one or more platters that rotate while in use.

Diskette A flexible disk, used principally on minicomputers and terminals for the storage of programs and information.

Display A device that visually represents information (e.g., CRT) on a screen.

Disposal Discarding and purging information.

Distributed access Communications with a computer from remote locations via communications lines and terminal devices.

Distributed computing Sharing of the computing among computers at several locations.

Distributed data entry Distributing the data entry effort among data entry facilities at several locations.

Distribution Message carrying, mail handling, electronic transmission, and traveling.

Documentation The management of documents which may include the actions of identifying, acquiring, processing, storing, and disseminating them.

Dot matrix The printing of characters using a series of dots. Characters formed are not always complete because they are not fully formed.

Downtime That portion of time when equipment is malfunctioning and therefore unavailable to its users.

Edit To prepare information for a letter or text operation. Editing may include the rearrangement or addition of information, the deletion of unwanted information, format control, code conversion, and the application of standard processes, such as pagination.

Electronic mail An electronic system utilizing communications lines, computers, CRT, and hardcopy terminals to transmit documents and messages to addressees.

Environmental issues Office automation concerns centering around the generation of noise, heat, particular matter from printers, radiation from terminals, lighting, furniture placement, regulatory restrictions and demographics.

Ergonomics Those aspects of an environment that focus on health, safety, and human comfort issues such as posture, sight, hearing, heat, noise, humidity, color, and space.

Expansion Copying (reprographics), printing, and micrographics (microfilming).

Facsimile (FAX) A system for the transmission of images. The image is scanned at the transmitter, reconstructed at the receiving station, and duplicated on some form of paper.

Feasibility study A preliminary investigation to determine whether a task or project appears reasonable and will produce desired results to meet certain objectives.

Field In a record, a specified area used for a particular category of data, for example, a group of card columns in which a wage rate is recorded.

File A set of related records treated as a unit. In electronic filing, a file could consist of a set of letters or other documents on a particular subject.

File protection Prevention of the destruction of information recorded in storage.

Fixed assets Items utilized in conducting ongoing business such as building and equipment.

Floppy disk A small mylar storage device resembling a phonograph record.

Flowchart A graphical representation for the definition, analysis or method of solution of a problem, in which symbols are used to represent operations, data, flow, equipment, etc.

Font The type style of a typewriter, word processor, printer, or display.

FORTRAN (Formula Translation) A programming language primarily used to express computer programs by arithmetic formulas.

Frame A screenful of information on a video screen.

Friction feed The paper feed mechanism on printers by which friction guides the paper over the plates.

Full duplex A communications channel that can send and receive simultaneously.

General purpose computer A computer designed to operate on a wide variety of problems.

Half spacing A feature that permits text to be displayed or printed in half space increments.

Hard copy Output from a word or data processor in printer form.

Hardware Physical equipment used in data processing, word processing, or other systems, as opposed to the computer programs, procedures, rules, and associated documentation.

Impact printer Prints characters by mechanical means such as a type ball.

Inflation An increase of price levels over time that is generally caused by an increase in the money supply and readily available credit.

Information The meaning a human assigns to text, data, images, and sounds by means of the known conventions used in their representation.

Information categories Identifications of the most commonly utilized processes, procedures, and techniques used within the corporations for the creation, capture, preparation, distribution, storage, retrieval, and destruction of documents and other forms of information.

Information resource management The managing and developing of information systems telecommunications systems and office systems in a planned and controlled manner. This requires the acceptance of information as a commodity or asset in the broadest sense.

Ink jet Printing technique by which characters are sprayed onto the paper through a character template.

Input buffer Temporary storage used to permit data to be stored until it can be processed.

Inquiry A request for information from storage, for example, a request for the number of employees hired during the past year or a statement to initiate a search of library contents.

Intelligent copier Copier with a built-in microcomputer that can copy, collate, and communicate with word and data processors.

Interactive graphics A technology (hardware and software) that is on-line to a computer and has the ability to produce many kinds of graphics and pictures in color or black and white. In this environment, the user typically has the ability to interact with the system and can manipulate the graphics.

Interactive Those computer or word processing systems that users input into and receive immediate feedback from.

Interface The part of a system that connects with another system component, or that part of the system with which the user comes in contact.

Internal assessment A general review program with emphasis on the establishment of an equipment and services inventory and detailed quantifications and qualifications of how management, professional, technical, clerical, and secretarial staff utilize their time.

Job classifications Segmentation of employment levels for reference purposes. Management categories include: chief executive officer, vice presidents, upper management (general managers and directors), middle management (managers), first-line management (supervisors), professionals, and technicians. Secretary and clerical classifications include: executive secretary, management secretary (general managers and directors), secretary (middle and first-line management), office worker, clerk-/typist, and other nonexempt staff.

Job enrichment Making a job more fulfilling by giving an individual more authority and independence.

Keyword Used to retrieve information in an electronic file.

Language (computer) A set of characters, conventions, and rules used for instructing a computer to perform certain operations.

Large scale integration (LSI) The integration of hundreds or thousands of functions and circuits on a single chip.

Learning curve The relationship of learning to the period of training or experience.

Library A collection of related files.

Line printer A printer that prints one line of text as a single unit.

Lines per minute (LPM) The number of lines a printer can print in one minute.

Magnetic tape A tape with a magnetizable surface layer on which information can be stored by magnetic recording.

Mainframe computer A medium to large general-purpose computer.

Main storage Computer-addressable storage from which instructions and other data can be loaded directly for subsequent execution by a computer program.

Manuals Reference materials that usually consist of policies, procedures, and memorandum reflecting management's statement of the methods on which a particular function should be established.

Matrix organization An organization that blends centralized and decentralized organizational philosophies.

Microcode The instructions that control certain functions of terminal systems and computers, also termed firmware.

Microprocessor One or more LSI chips that perform the functions of a central processing unit.

Microsecond (us) One millionth of a second.

Millisecond (ms) One thousandth of a second.

Minicomputer A digital computer supporting peripherals such as printers, video displays, disks, and magnetic tape. Usually presented between microcomputers and mainframes.

Modem Equipment that "modulates" and "demodulates" digital information into an analog signal and vice versa.

Multiprogramming A mode of operation that provides for the interleaved execution of two or more computer programs by a single processor.

Multitasking Multiprogramming that provides for the concurrent performance, or interleaved execution, of two or more tasks.

Murphy's Law A belief that whenever something can go wrong, it will.

Nanosecond (nsec) One billionth of a second.

Networking The interconnection of computers and terminals by means of communication lines.

Office automation The use of technology in the office to perform functions previously performed by office workers.

Office systems The use of technology by office workers to augment their performance on office functions.

Offline Pertaining to the operation of a functional unit without the continual control of a computer.

Optical character recognition (OCR) The reading and identification of characters optically under computer control.

Online Pertaining to the operation of a functional unit that is under the continual control of a computer.

Operating system Software that controls the execution of computer programs that provides scheduling, debugging, input-/output control, accounting, compilation, storage assignment, information management, and related services.

Organization charts Representations of the structure of the corporation. Generally reflects senior operating management, although similar charts are also available for departmental and division management.

PABX Private automatic branch exchange.

Page A frame or screenful of data.

Personal computer A wide range of small computers used in the home, or office or factory.

Picosecond One trillionth of a second.

Pilot project A prototype project established to test an idea, concept, technology, or product.

Pitch The spacing between characters printed on a horizontal line. Spacing is typically 10 pitch (10 characters to the inch) or 12 pitch (12 characters to the inch).

Planning A structured process that requires the establishment of short, intermediate, and long-range goals, objectives, and strategies. These may be quantitative and/or qualitative.

Platen The round cylinder or roller that holds and moves the paper in a printer or typewriter.

Preparation Entry into a keyboard.

Private (automatic) branch exchange PBX (PABX) A telephone switching computer used within a company facility to route telephone calls and other forms of information.

Process A systematic sequence of operations to produce a specified result.

Program A series of actions designed to achieve a certain result.

Programmer A person who designs, writes, and tests computer programs.

Programming The designing, writing, and testing of computer programs.

Programming language An artificial language established for expressing computer programs.

Project A well-defined activity in terms of duration, "deliverables," costs, schedules and responsibilities.

Proportional spacing The allocation of specific line spacing for each printed or displayed character.

Protocol The communication "language" between communication devices.

Queue A line or list formed by items in a system waiting for service; for example, tasks to be performed or messages to be transmitted in message switching system.

Record A collection of related data or words, treated as a unit; for example, in stock control, each invoice could constitute a record.

Regulatory reporting Requirements, established by law, under which corporations must report various types of operating detail relative to financial, tax, employee, shareholder, and other information.

Requisites A set of coordinating and control techniques used to monitor an office automation program, such as standards, progress reports, plans, and steering committees.

Return on investment (ROI) Conceptually, ROI is the ratio of the income to the investment cost. Because there are many different ways of measuring income and investment, ROI is used here as a generic term.

Right justification The printing or displaying of information with the last character of each line aligning in a single column (as in this book).

Satellite communications Use of orbital satellites to transmit information (voice, data, video, graphic) between two or more locations.

Search, storage, and retrieval Indexing, storing, searching, and finding information.

Security Prevention of access to or use of information or programs without authorization.

Soft copy Output from a word or data processor displayed on a CRT.

Software Computer programs, procedures, rules, and associated documentation concerned with the operation of a data processing system.

Span of control The exercise of management over a category of staff. Generally used to indicate the amount of responsibility an individual or group can exercise and still retain proper levels of guidance and control over the function arranged.

State of the art The most recent level of technology.

Storage and retrieval Business system components that include such technologies as electronic storage/retrieval systems and automated micrographic systems.

Study team Multidisciplinary group established to analyze all aspects of administrative activities including business systems, planning, statistics, human factors, and other relevant areas.

Survey tools Questionnaires, logs, survey tools, and other techniques utilized to determine quantitative and qualitative information about office activities.

System In data processing, a collection of men, machines, and methods organized to accomplish a set of specific functions.

Systems analysis The analysis of an activity to determine precisely what must be accomplished and how to accomplish it.

Task A basic unit of work to be accomplished.

Teleprocessing A form of information handling in which a data processing system utilizes commmunication facilities for data transmissions.

Terminal A device, usually equipped with a keyboard and some kind of printer or display, capable of sending and receiving information over a communication line.

Text editor A system that captures keystrokes on a magnetic medium so that they can be played back and changed without rekeyboarding the entire document.

Traditional The way things were.

Transitional The way things are changing.

Transformational The way things have changed.

Turnaround time The amount of time necessary to complete a process, such as the preparation of a document (typing, proofing, editing, and correcting) or the response to a computer inquiry.

Turnkey system The hardware and software ready for installation and operation as a user's system.

Universal product code scanner A device that incorporates photographic processes (holography) with laser and computer techniques to scan information from consumer items in stores and translate the information into price and inventory statistics.

User An individual or a department in an organization that supplies input to and/or uses output from information systems—
or one that uses the systems itself.

Word processing The keyboarding, editing, and printing of
documents on a system that includes hardware and, usually,
software.

Index

321